UNDERSTANDING
BIRD BEHAVIOUR

A BIRDWATCHER'S GUIDE

UNDERSTANDING
BIRD BEHAVIOUR

A BIRDWATCHER'S GUIDE

Stephen Moss

NEW
HOLLAND

First published in 2003 by New Holland Publishers (UK) Ltd
London • Cape Town • Sydney • Auckland

Garfield House, 86–88 Edgware Road, London W2 2EA,
United Kingdom www.newhollandpublishers.com

80 McKenzie Street, Cape Town 8001, South Africa

Level 1/Unit 4, 14 Aquatic Drive, Frenchs Forest, NSW 2086,
Australia

218 Lake Road, Northcote, Auckland, New Zealand

10 9 8 7 6 5 4 3 2 1

ISBN 1 84330 151 2

Publishing Manager: Jo Hemmings
Project Editors: Lorna Sharrock, Gilly Cameron Cooper
Editorial Assistant: Gareth Jones
Cover Design: Alan Marshall and Gulen Shevki
Design: Warrender Grant
Copy Editor: Mike Brown
Index: Janet Dudley
Production: Joan Woodroffe

Reproduction by Modern Age Repro Co., Hong Kong
Printed and bound in Malaysia by Times Offset (M) Sdn Bhd

Cover photos: Osprey; Back cover: Gannets; Front flap: Barn Owls
Page 3: Magpie
Page 6: Skylark, Kingfisher, Gannets
Page 7: Barn Owl, Robin, Puffin

THE WILDLIFE TRUSTS

The Wildlife Trusts partnership is the UK's leading voluntary organization working, since 1912, in all areas of nature conservation. We are fortunate to have the support of more than 411,000 members – people who care about British wildlife.

We protect wildlife for the future by managing almost 2,500 nature reserves across the United Kingdom. The habitats on the reserves range from wetlands, peat bogs, and coastal and marine environments, to heaths, woodlands and meadows. Many are home to birds such as the rare Bittern and Manx Shearwater, Puffins, Skylarks or Barn Owls. While birdwatching in such places, you may well catch sight of other wildlife – a basking shark, perhaps, or a red squirrel. These and many other animals are protected by Wildlife Trusts across the UK.

The Wildlife Trusts work with the general public, companies, planners and Government to raise awareness of species and habitats that are increasingly at risk. Importantly, we encourage people to experience wildlife for themselves. We believe that a deeper appreciation for nature conservation can start with a book such as this one by Stephen Moss. We need more people to understand and value the birds and other wildlife that are to be found in our countryside.

Understanding Bird Behaviour introduces the reader to the habitats and characteristics of birds in life – how and why birds feed, preen, and react with others of their kind. The author compares such behaviour as courtship, fledging, flight and migration among many different species, and investigates the instincts and circumstances that trigger these behaviours.

Few realize just how endangered much of our British wildlife is. In recent years, once-common bird species such as the sparrow and Starling have declined, mainly due to the demands that modern human living has placed on habitats.

The Wildlife Trusts believe, however, that it is not too late. Much can still be done to reverse the losses of the past, and we all have a part to play in making this happen. One way is to contact your local Wildlife Trust for information on wildlife activities such as dawn chorus events and volunteering opportunities, and on nature reserves. Help us to protect wildlife for the future and become a member today! Please complete the form attached, or phone The Wildlife Trusts on 0870 0367 711. Click on to www.wildlifetrusts.org for further information. The Wildlife Trusts is a registered charity (number 207238).

We hope that, with the help of this book, you have fun learning more about birds and their behaviour!

CONTENTS

INTRODUCTION

Studying bird behaviour is one of the most fascinating and potentially rewarding aspects of watching birds. But where do you start? At first, understanding what birds are doing, and more importantly why, can be confusing, especially if you are a newcomer to birding. Is the aspect of behaviour you are witnessing a normal part of daily life, or something unusual? Will your presence disturb the bird and force it to behave out of character? And how do you interpret some new or different aspect of behaviour you have not witnessed before?

This book will provide some of the answers to these, and many other questions. Its purpose is three-fold:

• To provide an introduction to the various different forms and aspects of bird behaviour, categorized by subject area.

Fulmars live up to the origin of their name, which means 'foul gull', by spitting a foul-smelling oil at intruders if they approach too close.

The Blackbird is one of our best-known and best-loved songbirds, thanks to the purity and tone of its song.

- To indicate specific types of behaviour that are characteristic of certain species or family groups.

- To be a work of reference. Using the index, you can look up either a particular species or a specific aspect of bird behaviour that you want to find out more about.

The book is divided into two parts, each of which can be read either independently, or by cross-referring between them.

Part 1 Types of Bird Behaviour covers the various types of behaviour applicable to birds in general, such as flight, courtship, predation and migration. These are loosely grouped under the chapter headings of: Movement; Feeding; Breeding; Migration and Navigation; Distribution and Range; Life and Death. This provides a quick and easy reference to specific behaviours.

Part 2 Species Behaviour is arranged by families or similar groups of birds such as seabirds. This includes the 200 or so species that you are likely to encounter in

Like most newly fledged birds, juvenile Blue Tits are sociable, staying together with their parents and siblings for a while after leaving the nest.

Britain and Northern Europe, some common, others rare, with details of behaviour common to that particular species or group. This enables you to look up individual species and get some insight into their behaviour, though for reasons of space this cannot possibly be comprehensive. If you are interested in following up the behaviour of a particular species or family, details of suitable works can be found in the Further Reading section at the back of the book.

During the past couple of decades, birding has focused predominantly on the two aspects of identification and rarities. While these are both fascinating and important, they have absorbed the attention of birders' minds at the expense of another vital aspect of understanding and enjoying birds – the study of their behaviour.

Recently, twitching (the hobby of zooming around the country to see a rare bird), has gradually declined. Simultaneously, there has developed a more enlightened attitude towards the pastime of birding as a way of getting back in touch with nature. The study of bird behaviour is therefore due a renaissance, and this book is a small contribution towards encouraging all birders to take an interest in it. To me, recognizing and understanding how different species behave is by far the most interesting aspect of birdwatching of all.

A winter flock of Snow Buntings will brighten the gloomiest of days. White flashes from winter males really do look like a flurry of snow.

The arrival of flocks of Redwing is one of the sure signs that winter is just around the corner.

1 TYPES OF BIRD BEHAVIOUR

The first half of this book deals with the different types of bird behaviour. For the sake of convenience, these are divided into six chapters, each of which deals with a range of related aspects of behaviour:

• MOVEMENT: Feathers and flight; Swimming and diving; Walking and running; Flocking behaviour; Roosting and sleeping.

• FEEDING: Food types and feeding methods; Birds of prey; Specialist feeders; Obtaining water.

• BREEDING: Timing of breeding; Territory and birdsong; Courtship, display and mating; Nest building; Egg laying and incubation; Parental care and fledging; Hybridization and unusual breeding behaviour.

• MIGRATION AND NAVIGATION: Why do birds migrate? How do birds navigate? Migration strategies; Unusual migration.

• DISTRIBUTION AND RANGE: Habitats: their influence on behaviour; Range and distribution.

• LIFE AND DEATH: Moulting and plumage; Bathing, preening and feather care; Sight, hearing and smell; Excreting waste; Temperature regulation; Birds and weather; Disease and death.

To find information about particular species you may also want to refer to the second half of the book, which deals with behaviour on a family-by-family basis.

MOVEMENT

FEATHERS AND FLIGHT

One of the characteristics that separates birds from most other animals (apart, of course, from insects and bats) is their ability to fly. They are able to do so because of their unique body structure: a light, hollow skeleton supporting feathers, which enables them to take off and stay airborne. That said, birds have developed many different ways to fly, including soaring, gliding and flapping, with the help of the wind and air currents such as thermals.

Birds are able to fly because, over many millions of years, their bodies have undergone particular adaptations. Most important of these was the evolution of feathers – light, versatile structures, probably evolved from reptilian scales, that are unique to birds. Flight feathers, in the bird's wing and tail, are stiff and long, enabling birds to gain and maintain lift and to manoeuvre themselves through the complexities of air currents once aloft. In addition, a bird's skeleton is also highly adapted to flight. Light, hollow bones carry the minimum of extra weight, with the result that birds are by far the lightest of all animals in relation to their size. The classic flight

Birds, especially seabirds, are the true masters of the air. This Great Shearwater uses air currents to glide across the surface of the ocean.

Day-flying raptors, such as this Common Buzzard, have broad wings, which enable them to gain lift rapidly by using warm air currents. Their wing shape may change, depending on whether they want to soar to stay aloft in one place, or to glide to travel distances in search of prey.

mechanism is flapping – moving the wings up and down to gain lift. It is generally used for short, direct flights (for example, a songbird moving from tree to tree), as it uses up a lot of energy. Once aloft, or when travelling for any distance, most birds prefer to use less energy-expensive methods of moving through the air, such as gliding and soaring. Seabirds such as **albatrosses** and **shearwaters** are the world's greatest gliders, taking advantage of updraughts from the ocean surface to maintain their position just above the waves. In this way, they can move forward using the minimum of energy, hardly flapping their wings, for many hours on end. Raptors, such as **hawks**, **buzzards** and **eagles,** also use gliding flight. They place their wings

so as to reduce surface area and allow rapid forward movement while maintaining lift. Raptors also spend much of the time soaring, a flight style that is particularly common among large, heavy birds. When soaring, a bird such as a **Buzzard** opens its wings as far as possible, then takes advantage of thermal air currents of rising hot air to gain height. Soaring is an energy-efficient way to stay aloft, as by maximizing its wing area, the bird uses less energy. Once aloft, it can circle around for some time, again using the minimum of energy. Soaring is generally used to maintain altitude rather than to move any great distance.

To see the difference between soaring and gliding, watch a **Sparrowhawk** as it soars overhead on broad, outstretched wings; then note how it changes its wing angle, narrows the wings and glides rapidly across the sky, appearing quite different in shape from before. Gliding on narrowed wings enables the birds to cover distances more quickly and efficiently than they would with open wings.

Scientists have compared flight with other locomotion mechanisms such as swimming or walking, and revealed that it is extraordinarily efficient in comparison. For example, the fastest sprinter in the world covers the ground at about 5 body-lengths per second, while even the world's fastest land animal, the cheetah, can only manage 18 body-lengths per second. A flying bird, however, can reach up to 70 or even 80 body-lengths per second, a speed comparable with a jet aircraft. This not only allows birds to get to their destination quickly, but it also enables them to cover vast distances, especially on migration.

Sparrowhawks can stretch their wings out to soar, or glide high in the open sky. Short, rounded wings and a long tail make Sparrowhawks very agile, enabling them to manoeuvre quickly through the dense foliage of their woodland or garden habitat.

SWIMMING AND DIVING

Of course, not all birds spend the majority of time in the air. Many waterbirds spend most of their lives in an aquatic habitat, and have adapted their physiology and behaviour accordingly. At its simplest, this involves adapting flight techniques when swimming underwater – watch film of **auks** or **penguins** underwater and you will see what I mean. Their wings, which look pretty useless above the surface, are transformed into efficient 'propellers', enabling them to cover distance and manoeuvre themselves underwater.

Other birds stay mainly on the surface, or divide their time between land and water. Wildfowl, such as **ducks**, **geese** and **swans**, have adapted to a life spent on water by evolving a number of features, such as fully webbed feet and oil glands

Gannets hunt by flying high above the surface of the sea, then folding back their wings and plummeting at great speed into the water below to catch unwary fish.

with which they waterproof their feathers on a regular basis to keep them sleek and in good condition. Other waterbirds appear to have only partially adapted to their aquatic lifestyle. **Grebes** and **coots** have partly webbed feet only, perhaps because they rarely swim long distances. **Cormorants** do not have waterproof feathers, so have to stand for long periods drying their wings in their characteristic pose.

Waterbirds from unrelated groups often show very similar features and superficial appearance. **Coots** and **moorhens** for example, look more like ducks than the other members of their family, rails and crakes. This is due to a process called convergent evolution, in which external factors dictate the morphology of a bird or other organism, leading it to superficially resemble other creatures that share the same habitat and lifestyle.

Diving is another feature shown by different, unrelated groups of birds, including **divers**, **grebes**, many kinds of **duck**, and seabirds such as **auks**. Again, these species share similar features: a long, streamlined body, webbed or partially webbed feet for underwater propulsion, and legs situated well back on their body. This means that for some specialized birds, such as divers, getting about on land proves very difficult indeed.

Razorbills, like other members of the auk family, have short, powerful wings that enable them to swim effectively underwater. In flight, they must beat them rapidly in order to stay airborne.

WALKING AND RUNNING

For many land birds, the best way to travel short distances without using too much precious energy is also the simplest: to walk, hop or run. This is especially true of ground-feeding birds, such as **finches**, **sparrows** and **thrushes**. Different groups use different methods. For example, tree-dwelling birds, such as finches, tend to hop when on the ground, while **larks** and **pipits**, which spend much more of their time on the ground itself, will walk or run.

Other groups, such as **game-birds**, **rails** and **crakes**, spend the vast majority of their time on the ground and are well adapted to running and walking, often hiding in dense cover and only using flight as a last resort.

Game-birds, such as these Red-legged Partridges, often prefer to run rather than fly, as it makes them less vulnerable to predators.

FLOCKING BEHAVIOUR

Birds are, by and large, sociable creatures, and many different species gather in flocks, either occasionally or on a regular basis. Flocks occur for various reasons: to maximize the chances of finding food, to keep warm, or to avoid predators. It is important to understand that in all these cases the urge to flock must come from each individual, and that the advantages of joining a flock must outweigh the disadvantages for that individual, rather than the group as a whole. Flocking also occurs as a by-product of the concentration of food supplies. For example, a rubbish tip will attract all the local **gulls**, which will fiercely compete for the available food.

Flocking is much more widespread outside the breeding season, for the simple reason that, during the nesting period, most birds are in pair bonds with one or more members of the opposite sex, and their duties, such as incubation and feeding the young, preclude joining others in flocks. Once breeding is over, songbirds gather in family parties or loose flocks of a dozen or so. While food is plentiful and there is sufficient cover to avoid predators, there is no pressing need to form larger groups.

Birds of many different species come together in flocks, especially in winter, when food resources are scarce, and they are more vulnerable to predators.

In winter, factors such as reduced amount of daylight in which to find food motivate individuals to seek out others. From late autumn you often see flocks of **tits** travelling through woods in search of food, keeping together by uttering brief contact calls. Flocks like this are more efficient at finding food and avoiding predation, thus making it beneficial for each individual bird to join. Seed-eaters such as **finches** and **sparrows** do the same.

Dusk in winter is a great time to watch vast flocks of birds, such as these Starlings at Brighton's West Pier, gathering to roost.

Waders and **wildfowl** also form large flocks outside the breeding season. In most cases this is because their food is plentiful in specific areas such as mudflats or estuaries. Flocking has the added advantage of making them less vulnerable to predators such as birds of prey, which may be confused by the swirling mass of birds and therefore unable to focus on a particular individual. It takes one bird only to spot a hunting **Peregrine**, and give an alarm call, and the whole flock can take to the air, wheeling from side to side to confuse the attacker.

A few species form large flocks only occasionally, such as a feeding frenzy of a variety of species of **seabirds** behind a trawler, or a gathering of **Magpies** in an area with a high population of the species. Again, to do so must confer an advantage, however small, on each individual – otherwise they would disperse and search for food on their own.

The habit of forming flocks can be a great help to birdwatchers, especially in woodland or farmland habitats. At first, it can sometimes seem as if there are no birds present at all, but watch for movement, and listen for the contact calls made by members of a flock, and sooner or later you'll come across the flock itself.

To observe flocking at its most spectacular, you will need to visit certain specific habitats, such as estuaries, traditionally managed farmland, or a specific site for a **geese** or **Starling** roost. Make sure you check details such as tide times or the time the sun sets, as these can be critical: arrive too early and the birds will still be out feeding in the marshes or fields; arrive too late and you will have missed the spectacle.

TOP TIP

Flocking
Birds in flocks appear to move almost as a single organism, twisting and turning in perfect synchrony. In fact, each bird takes its cue from the birds around it, following their movement almost instantaneously, and giving the impression of one mind at work. In normal flocks there is no single leader. Migratory flocks are different: experienced birds take turns to lead.

Autumn and winter are the best times to look for tits – they can usually be seen with several different species, which keep together by uttering frequent, high-pitched calls.

ROOSTING AND SLEEPING

Flocking is often the prelude to roosting: either for the night, as in most groups such as **gulls**, **crows** and **Starlings**, or, in the case of **waders**, because the high tide has temporarily covered up the food supply. Roosting confers many of the same advantages as flocking, notably warmth during cold nights and safety in numbers against predators.

The vast majority of the world's birds follow a typical circadian cycle, which is linked to the 24-hour rhythm of the day. Thus, as evening approaches, diurnal species such as the **Starling** leave their feeding areas and gather together at a single roosting place, where they spend the hours of darkness. Starling roosts have been known to contain several million birds, making them an awe-inspiring sight, especially as the birds wheel about the sky in vast flocks just before and after sunset. Whether or not

Jackdaws coming in to roost can put on an amazing aerial display, as they drop out of the sky to land on their chosen site.

20

Flocks of magpies can be found roosting in woodlands and hedgerows – and have given rise to a number of country rhymes, such as the well-known, 'One for sorrow, two for joy…'

this behaviour has any hidden purpose, such as communicating the whereabouts of food resources, is not clear.

The safety factor is clearly a prime motivation for communal roosting. By joining other birds, an individual has a far greater chance of avoiding predators. Warmth – particularly crucial during winter months – is another key factor. By huddling together, birds are better able to conserve valuable body heat.

Birds that do not usually form flocks during daylight hours often roost together at night. For example, **Pied Wagtails**, which tend to feed singly or in pairs during the day, come together in roosts of 50 or more birds at night. **Wrens**, normally pugnacious and solitary birds, will also gather together, especially during very cold winter weather. Up to several dozen birds have been found huddling together in a single nestbox.

In bad winter weather, Wrens will often gather to roost in nestboxes, in order to gain extra warmth from their companions.

A characteristic prelude to roosting is a flight line, in which hundreds, sometimes even thousands of individuals can be watched as they make their way from their daytime feeding areas to their roosting site. This is especially prevalent amongst **gulls**, which can be seen following strictly defined routes each morning and evening as they go to and

A daytime roost is often the best time to see nocturnal species, such as the elusive Long-eared Owl, though you must be careful not to disturb the sleeping birds.

from roosting and feeding areas, sometimes a distance of many miles. Other birds that regularly follow flight lines include **crows** and **thrushes**. Flight lines are simply the most efficent way to get from feeding to roosting areas; either the most direct route, or one following a landmark such as a valley or river.

Not all birds roost communally in large groups. Many songbirds choose to spend the night either on their own or in a loose group, though the same factors of safety, security and warmth are just as important. A small number of birds, such as **owls** and **nightjars**, are predominately nocturnal and roost by day. Owls tend to roost singly, and can sometimes be hard to locate. However, they are creatures of habit and will often occupy the same roosting place from day-to-day and even year-to-year, providing they are not disturbed, making them relatively easy to find once you know a regular site.

Once at a roost, birds often seem so noisy that you wonder how they can ever get any sleep. This may be because the human notion of eight hours of uninterrupted, deep sleep is quite different from the experience of most birds. Even when safely together in a roost, birds always need to be wary of potential predators, so they tend to 'cat-nap', often tucking their head under a wing and sleeping while standing up. Some birds, notably **Swifts**, appear to be able to sleep on the wing, but due to the difficulty of actually observing this phenomenon it has not been absolutely proved that they do so.

Not all birds sleep at night. Those dependent on the state of the tides for their feeding, especially waders such as **sandpipers** and **plovers**, must attune their bodily rhythms to tidal rather than

TOP TIP

Roosting
The best time to watch birds going to roost is during the hour or so before dusk, when the light is still good enough to identify the species involved. However, even after night falls, latecomers may still be coming along, so don't go home too early.

diurnal cycles, and can often be seen fast asleep during broad daylight. Conversely, they, and other species such as **ducks**, often feed by night, especially if there is a full moon to enable them to find their prey more easily.

Many coastal species of wader, such as the **Dunlin, Oystercatcher** and **Knot**, are able to feed only when the tide is out, exposing the food-rich mudflats of estuaries and saltmarshes. Therefore, as high tide approaches, the birds need to find a suitable place to roost.

Being in the right place at the right time when thousands of waders come to roost can be an unforgettable experience. Two or three hours before high tide, as water starts to cover the feeding-areas, large flocks begin to form. This is often the best time to observe them, as they wheel around the sky or move from place to place. As the tide finally reaches its height, the waders gather in vast numbers, either occupying the few small areas of mud left uncovered by the sea, or roosting on nearby non-tidal areas such as gravel-pits. Once settled, they rest and sleep for several hours, before the waters recede and they are able to feed once again.

Ducks such as this pair of Teal will feed throughout the day – and sometimes at night as well.

Feeding time for waders, such as these Knots and Oystercatchers, is dependent on tide cycles, rather than time of day. At high tide they gather in vast roosts to rest and sleep.

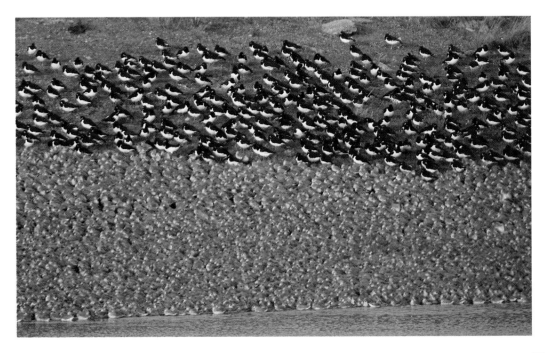

FEEDING

Thrushes, such as this
male Blackbird, love berries,
gorging themselves on the
soft fruit all day long.

FOOD TYPES AND FEEDING METHODS

One of the ways in which birds have adapted more widely than other animals is in their feeding methods. Between them, the world's birds eat almost anything, from shellfish to seeds, berries to bees, and nuts to nectar. Some even eat other birds! So it is hardly surprising that the various species and families have adapted to feed in many different ways, and, in doing so, have developed an extraordinary range of differently shaped bills.

Even within a single group, such as the songbirds, there is a vast array of different feeding methods. Seed-eaters, such as **finches, sparrows** and **buntings**, may prise their food from the heads of flowering plants, or simply pick up seeds from the ground. Some, like the **House Sparrow**, have all-purpose bills designed to enable them to take a wide variety of different seeds. Others, such as the **Goldfinch**, have highly specialized bills which have evolved to obtain seeds from plants such as the teasel that other finches find it impossible to get to. The most specialized seed-eaters of all, **crossbills**, have evolved an

TOP TIP

Feeding and flocking
Knowing how different birds feed can be a useful way to locate them, especially in winter, when many species form communal flocks to find food. Listen out for their contact calls, which are used to keep the flock together and alert others to a new source of food.

extraordinary and unique bill in which the tips of the mandibles cross over. This enables them to prise open pine cones and extract the seeds.

Insect-eaters, too, have evolved a wide range of feeding methods. Many **warblers** obtain their food by gleaning, that is working their way around the leaves, twigs and branches of a tree, and picking up tiny insects as they go. **Goldcrests** have developed this into a method known as 'hover-gleaning', in which they flutter like a hummingbird while seizing tiny morsels from the tips of leaves.

Flycatchers, as their name suggests, launch themselves into mid-air, grab a flying insect, then return to their perch to devour it. **Bee-eaters** do the same, knocking the bee to stun or kill it and removing the sting before consuming it whole. Meanwhile, flying hunters, such as **swallows**, **martins** and **swifts**, have broad, wide bills, which enable them to seize flying insects in mid-air.

Other songbirds are omnivorous, and have more generalized bill shapes and feeding methods. **Tits** can take seeds and insects, and feed their young on caterpillars. Members of the **crow** family, such as the **Jay** and **Magpie,** will eat a very wide variety of foods, and have all-purpose bills to do so. **Thrushes** also take

Spotted Flycatchers have evolved the ability to catch tiny insects, such as flies, in mid-air – a spectacular sight.

Some species, such as this Goldfinch, have specially adapted bills in order to feed on specific plants – in this case to obtain teasel seeds.

a variety of food, but many have long bills enabling them to catch earthworms, while they are also able to eat berries in winter. The ability to vary diet from season to season is a crucial factor in survival, especially for those birds that, although insect-eaters in spring and summer, stay put in northern latitudes for the winter. A good example is the **Blackcap**, which in winter will change its insectivorous diet to one of seeds and berries, enabling it to overwinter in Britain.

Many of these species have adapted very readily to food provided in artificial feeders by humans. Once this habit was confined to **tits** and **sparrows** on hanging feeders, and **Starlings**, **Robins** and a few other species on bird tables. But a revolution in the design of bird feeders, and the quality and variety of food, has enabled many more species to feed in this way. Today it is a common sight to see quite specialist feeders such as **woodpeckers** and **Nuthatch** on bird feeders.

Wildfowl, too, show a range of different feeding methods. Many species of **swan**, **geese**, and **ducks** such as **Wigeon** regularly graze on grass or other crops (often to the annoyance of farmers). Others, especially the so-called 'dabbling ducks', such as **Mallard**, **Teal** and **Pintail**, dabble for morsels of food on the surface of water, while, with its purpose-built bill, the

Unlike most other ducks, Wigeon feed primarily by grazing on low-level grass, using their specially adapted short bill to pluck individual strands.

Shoveler filters tiny organisms just below the water's surface. **Swans** dip their long necks into the water, while 'diving ducks' such as **Tufted Duck** and **Pochard** dive beneath the surface.

More than any other group of birds, **waders** show great adaptability when it comes to feeding methods, as shown by their extraordinary variety of bill size and shape. Many, including common species such as **Dunlin** and **Redshank**, have fairly standard bills, which probe beneath the mud or pick items off the surface. Others, including the **Curlew** and **Avocet**, have long, curved bills, which respectively probe for lugworms or sift the water for tiny invertebrates. Watching a variety of waders feeding is an object lesson in how evolution produces different body shapes from the same basic design.

Seabirds have also evolved a wide variety of ways to obtain food. Some, such as the **Gannet** and many species of **tern**, fly over the water, then plunge down to obtain food. Others, especially **auks**, swim on the surface, then dive down, sometimes hundreds of metres below. **Gulls** are generalists, either grabbing morsels from the surface or leaving the sea altogether to feed in fields or even rubbish dumps.

Like other waders, Knots have long legs in order to go into the water to find food from the mud below the surface. They feed on small marine invertebrates, including worms and molluscs.

BIRDS OF PREY

In a sense, any bird that feeds on another living creature is a predator. However, in general use we reserve the term 'birds of prey' for two unrelated, but superficially similar, groups: **day-flying raptors** and **owls**.

The first of these groups includes a wide range of different species with an equally wide range of hunting and feeding methods: **eagles**, **hawks** and **buzzards**, **harriers** and **kites**, **falcons**, and the fish-eating **Osprey**. They all share the characteristic traits of fierce talons to catch and grip their prey, and a sharp, hooked beak to tear it apart for feeding. However, they hunt and catch their prey in a variety of interesting and different ways.

Our commonest and best-known bird of prey, the **Kestrel**, often hunts by hovering motionlessly over a grassy verge, then plunging down to catch an unsuspecting vole. In contrast, the **Hobby** grabs insects such as dragonflies, or even small birds such as martins, in mid-air, while the **Peregrine** hunts by 'stooping' from a great height down onto its prey. **Sparrowhawks** use their

Kestrels have evolved the habit of hovering almost motionless over an area of short grass, in order to hunt their favourite prey – voles.

Little Owls are adaptable and fairly omnivorous feeders, catching their prey by diving down from a high position on a post or from the branch of a tree. They feed on insects (especially beetles), small mammals such as mice and voles, and small birds.

short wings and long tail to manoeuvre their way through dense woodland or garden foliage, ambushing their target, while larger raptors such as **eagles** and **buzzards** fly high above their territory, searching for food. Buzzards will search out carrion or rabbits, while eagles will look for hares or grouse. The **Osprey** hunts in a unique way for a raptor, plunging feet first into a lake to grab large fish in its talons.

Owls are mainly nocturnal, and tend to hunt by either watching and waiting for prey from a tree or post (**Tawny** and **Little Owls**), or gliding low over open ground, searching for the slightest movement of a small mammal (**Barn Owl**).

SPECIALIST FEEDERS

A few species have evolved highly specialized feeding methods, either to exploit a particular kind of prey (for example, **Bee-eaters** and bees), or to gain access to a food resource that would otherwise be denied to them. **Kingfishers**, for example, sit on perches above streams and ponds, then plunge below the surface to grab small fish, a behaviour trait not shared with any other British bird.

Kingfishers are the champion fishermen, able to plunge beneath the surface of a stream or river and seize their prey with a powerful, dagger-like bill.

Sometimes a bird may adopt feeding methods quite different from those of its relatives. For example, another aquatic bird, the **Dipper**, is related to wrens and thrushes, yet has evolved to exploit a very different environment from that of their terrestrial relations. Dippers live on fast-flowing streams and rivers, and perch on rocks before plunging right beneath the water surface to obtain small aquatic invertebrates.

Perhaps the most bizarre, yet highly effective, feeding method of all is known as kleptoparasitism, or, as it is often called, piratical behaviour. It is commonly practised by **skuas** and some species of **gull**, and

Great Skuas – also known by the Shetland dialect name of 'Bonxies' – are piratical hunters. They chase other seabirds and force them to give up their catch by harassing them. The targeted bird then panics and regurgitates its food.

involves one or more birds chasing an unfortunate victim (usually of another species, such as a small gull or tern), and forcing it to either drop, or in some cases regurgitate, its food, which the chasing bird then seizes and swallows. This behaviour is especially common in seabird colonies, where species, such as the **Arctic Skua** (or 'Parasitic Jaeger' as it is known in North America), make the lives of terns very difficult indeed. A similar opportunistic form of piracy is also practised by many species of the **crow** family, especially **Magpies**, which will take the opportunity to seize food from other bird species, or even from small mammals such as foxes, if they get the chance.

OBTAINING WATER

It is easily forgotten that, as well as feeding, birds also have to drink to maintain their water levels. In fact, some species, especially insect-eaters, obtain much of the moisture they require from their food, and may have to drink every one or two days only. However, seed-eaters have to compensate for the lack of moisture in their food by drinking at regular intervals. They do so either by visiting streams, ponds or puddles, or by coming to artificial sources of water, such as garden ponds and bird baths.

Birds are unable to swallow in the way we can, so need to drink in particular ways. These include sipping a drop of water into the bill, then tilting the head back to allow the liquid to pass down the throat. Others do a form of sucking, which involves vibrating the tongue to pump water back into the gullet. Drinking birds are very vulnerable to attack by a predator, so they are often extremely wary, taking only a sip or two before looking round to make sure they are still safe.

A few species, notably **swallows**, drink while on the wing, swooping down to the water and skimming the surface as they go. This has the added advantage of reducing the risk of being caught by a predator.

Seabirds, especially ocean-going species such as **shearwaters** and **petrels**, have to obtain water while at sea. They solve the problem by using salt glands to remove the excess salt from sea water, enabling them to stay away from land for many weeks or even months on end.

TOP TIP

Drinking
A good way to observe birds drinking is to find a small pool in the middle of a wood and sit quietly waiting for birds to appear. However, make sure you do so during a spell of dry weather, when there are fewer alternative sources of water for the birds to visit.

Grey Wagtails, like many small birds, obtain most of their water either through their insect food, or by sipping small amounts from a pond or stream.

31

BREEDING

TIMING OF BREEDING

Mute Swans have a reputation for being belligerent, often engaging in battles with intruders. Here, the territorial male is successfully fending off his rival, and defending his own territory, so that he can breed in peace.

In spring, the thoughts of young men turn to love – along with the rest of the animal kingdom – and birds are no exception. During a brief few months at the start of the year, they must make hay while the sun shines. To put it more specifically, they must breed. This is a complicated business. The male bird must stake out a territory, fight off his rival males, and find a mate (or mates). The two of them then engage in complex courtship rituals, build a nest, and mate. The female lays her eggs and incubates them (sometimes with the male's help). Once the eggs hatch, the real work begins, as the young constantly demand to be fed. Finally, the young fledge and leave the nest, though even then they may require parental care. No wonder many birds look tatty and exhausted by the time midsummer comes.

Not all birds breed at the same time of year, by any means. Resident birds such as the **Blue Tit** and **Blackbird** habitually breed a month earlier than migrants such as the **Willow Warbler** or **House Martin**, though this is not always the case. Some migrants, such as the **Chiffchaff**, arrive back in March and get down to breeding straight-away, while residents, such as the **Yellowhammer**, may not start until June. **Crossbills** may lay eggs in January, while the first brood of **Mallard** ducklings often appears well before Easter. On the other hand some species, especially **pigeons** and **doves**, appear to breed virtually all year round.

Other species may have several broods. The **Blackbird**, for example, can have up to five, and may be nesting almost continuously from March to August. Others may have failed to breed at their first attempt, and try again much later. I once saw a brood of **Great Crested Grebe** chicks hatch in mid-September and still survive the winter unscathed.

In recent years, the timing of the start of the breeding season has changed. Global warming means that the start of spring comes a week or more earlier in many parts of Britain, and the birds have reacted accordingly, laying eggs up to 2 weeks earlier than 30 years ago. Indeed, the discovery of this phenomenon, using data collected by generations of amateur birders, was one of the first pieces of empirical evidence that global warming was more than just a flash in the pan.

With the recent onset of mild winters, some common species may even be tempted to begin nest-building before Christmas, though a cold spell would soon put a stop to this.

TERRITORY AND BIRDSONG

A spell of mild weather in February, and our woods and gardens are suddenly filled with the sounds of spring. **Song Thrushes** sing their repetitive tune from roof tops. **Great Tits** bounce around the blossom, calling 'tea-cher, tea-cher'. Deep in the undergrowth, tiny **Wrens** explode into song. Late winter snow and ice may bring a temporary halt to

Although these young Wrens are fledged and about to leave the nest, they will depend on their parents for food for a little while yet.

Song Thrushes sing their characteristic repetitive, tuneful song from the high branches of a tree or bush. By doing so, they make sure rival males and females can hear them, even at a distance.

Robins are one of our best-known and best-loved songsters, possibly because of their unusual habit of singing almost all the year round.

the chorus, but whatever appearances suggest, spring has arrived, at least for the birds. To paraphrase the 1950s' pop song, why do birds sing in the first place?

Broadly speaking, birdsong has two main functions: to defend a territory and attract a mate. Male birds generally arrive back on their breeding grounds earlier than their mates, and spend the first few days moving around their new territory, establishing the boundaries and advertising their presence to all-comers – especially rival males of the same species.

Some species, notably the **Robin**, sing during the autumn and winter, as unlike most songbirds, they hold winter territories. On a cold, dull, wintery day, theirs may be the only sound to be heard, but on a bright, sunny day in February, all kinds of other singers join the chorus, in anticipation of the breeding season.

The second, equally important function of birdsong is to attract a mate. As the females leave their winter quarters and return to their breeding habitats, males are in desperate competition, for those that fail to breed run the risk of dying before they can pass on their genes. This is especially true for many songbirds, whose life expectancy is only a year or so.

One odd but noticeable thing about birdsong is that in general, the best singers have the dullest plumages. The **Robin** is a notable exception, but how about the **Blackbird**, **Nightingale** and **Marsh Warbler**? The explanation is obvious when you think about it: birds with brightly coloured plumages do not need to sing a complex song to attract a mate, while those that are a dull brown or black do.

Along with food and habitat, birdsong is one of the most important factors in the ecological isolation of different species. This was first discovered by the great 18th-century naturalist Gilbert White, who managed to distinguish between the three species of 'willow-wrens' (**Chiffchaff**, **Willow** and **Wood Warblers**), by listening to their distinctive songs.

COURTSHIP, DISPLAY AND MATING

Once a singing male has attracted a mate, the serious business of courtship and pairing gets underway. After feeding, courtship is probably the most important aspect of bird behaviour. It takes all kinds of forms and, far from being confined to what we think of as spring, can begin as early as January.

Great Crested Grebes have one of the most spectacular of all courtship displays. The two birds face each other in the water like a pair of narcissistic teenage lovers, rubbing their bills together and shaking their heads in a ritualized pantomime to cement the pair bond. If you're really lucky, you may even see

TOP TIP

Birdsong at dusk
The dawn chorus is justly celebrated, but don't forget the 'dusk chorus'. On a fine evening in spring or early summer there is another spell of birdsong, less intense but arguably just as enjoyable as the early morning equivalent.

This pair of Great Crested Grebes are performing the famous 'penguin dance' as part of their courtship display.

Feral (or domestic) Pigeons perform a variety of fascinating displays as part of their courtship and pair-bonding, which eventually leads to mating.

This female Avocet (left) is leaning forward in a submissive posture, indicating that she is ready for the male to mate with her.

them go on to perform the memorable 'penguin dance', during which both birds gather water weed in their bills, then appear to stand up in the water, frantically paddling with their legs while waving the weed in each other's faces.

One of the easiest courtship rituals to observe is that of the **Feral Pigeon**. The male puffs himself up like a prize-fighter, then performs a little dance around the female, who generally looks singularly unimpressed by all the fuss. Finally, after much bowing and scraping, the male will attempt to copulate with the female,

though as often as not she will foil his advances. Another wonderful display is performed by a stunningly beautiful bird, the **Avocet**. The male approaches the female tentatively, dipping his upcurved bill into the water as if attempting to feed. Then, when he judges that she is receptive to his advances, he leaps onto her back, mates in a second or two, and jumps off. This is followed by an extraordinary little ritual in which he runs forward, away from her, pecking at the water, while she carries on feeding.

All this ritual has a serious purpose, of course. The male that can most impress the female is the one that will get the chance to mate and reproduce. And the female is under pressure too. She must choose the healthiest-looking male, to increase her chances of producing a long line of descendants. Courtship rituals may look like a bit of fun, but to the birds themselves, they truly are a matter of life or death.

Mating is usually brief, and may take place a number of times over a period of several days. The female usually adopts a submissive pose, allowing the male to mount her and introduce his sperm into her cloaca.

TOP TIP

Nesting material
In early spring, watch for birds collecting and carrying nesting material in your garden or the local park. You can help them out by providing useful material, such as straw and human hair, which the birds will use to build or line their nest.

NEST BUILDING

Once the courtship ritual is over, and the pair-bond is strong, nest building usually begins. Actually to say 'building' is not always accurate, as many species simply find a place to nest and deposit their egg there. This is especially true of colonial species such as many seabirds. **Guillemots** and **Razorbills**, for example, will simply lay their single egg in a depression on a cliff shelf. As the egg is pear-shaped it will roll in a very tight

Robins build their nests in some extraordinary places, including toilet-cisterns, watering cans and even under the bonnet of a car. This bird has chosen a more conventional spot.

Puffins nest out of the way of predators in rabbit burrows, which they line with some dry grass and other soft material.

circle, but does not fall off. Other birds, such as **Sand Martins, Kingfishers, woodpeckers** and **Starlings**, lay their eggs inside a hole in a sandbank or tree. This requires the minimum amount of 'building' effort. For most birds, however, a nest is required in order to keep the eggs safe and allow them to be incubated. Nests come in a whole range of shapes and sizes, and also vary greatly in the amount of care and attention that is taken to make them.

The 'classic' songbird nest is that made by species such as the **Blackbird** or **Song Thrush**: a neat cup of woven grass or small twigs lined with mud. Larger birds, such as the **Wood Pigeon**, make a much tattier-looking nest from twigs; it often looks so flimsy that you can see the eggs from beneath. One of the more ornate and complex structures is that built by the **Long-tailed Tit**. It comprises a small ball of hair, moss and feathers (up to 2,000 in a single nest), held together by lichen. It is this amazing nest that gave the species the old country name of 'bumbarrel'.

Grey Herons are one of the earliest of all our birds to start nesting. They begin to breed in late January, in large colonies known as heronries.

Other small birds, such as the **warblers**, often nest on the ground, or build their nest in the fork of a tree. **Goldcrests** hang their tiny nest and its precious contents on the end of a twig, the structure being so light it manages to stay in place.

Waterbirds, such as **Coot** and **grebes**, build floating nests out of aquatic vegetation; in the case of the **Great Crested Grebe**, once the eggs are laid the adults use this to cover them up when they search for food.

The largest nest of any British bird belongs to the **Golden Eagle**. It is a bulky structure of twigs up to 3 metres across and 5 metres deep (9 x 16 feet). One of the smallest of nests is that made by the **Wren**, but there is a catch: the male must make several nests before the fussy female is satisfied. She will choose only the best one in which to lay her eggs.

Long-tailed Tits make an extraordinary nest shaped like a small barrel, with a tiny entrance hole for the birds to go in and out.

TOP TIP

Hatching chicks
Once the eggs have hatched, you may notice a change in the behaviour of the parents. Both are now busily searching for food to take back to their hungry youngsters. Even after the chicks have fledged they may still be fed by the adult birds.

EGG LAYING AND INCUBATION

Once the nest is built, the female can lay her eggs. These may be anything from 1 (most **seabirds**), 3–5 (many **waders**), 4–12 (**songbirds** and **wildfowl**) to as many as 20 or more eggs (**game-birds** such as the **Pheasant**). These are usually laid one per day, with incubation generally beginning the day after the final egg is laid and the clutch is complete.

This is a risky time for the birds, as there are many predators, such as **Jays**, **Magpies** and squirrels, for whom a clutch of eggs makes a tasty meal. The weather can play a part, too, especially

Magpies are well known egg thieves – always on the look out for a chance to steal another bird's eggs. Magpies' own nests are loose, untidy and built from sticks, often in the fork of a tree.

a cold snap, or heavy rain, which can cool the eggs and make them infertile. So incubation is a full-time job for most birds. It is either carried out by the female alone (as in many lekking species such as the **Ruff** and **Black Grouse**), by the female with the help of the male bringing food (most **songbirds**), or by both sexes (most **seabirds**). In two rare British breeding species, the **Dotterel** and the **Red-necked Phalarope**, the roles are reversed, and it is the male that incubates the eggs while the female goes off on her own (see Aberrant Breeding Behaviours).

The period of incubation varies greatly. This can be from just 11 days for the **Lesser Whitethroat** and **Skylark**, to an astonishing 54 days in the case of the **Manx Shearwater**, and even longer for some **Fulmars**. For most songbirds, the usual incubation period is around two weeks; ducks, waders and game-birds about three to four weeks; and birds of prey, four to seven weeks. **Seabirds** have the longest incubation periods, of between four and almost eight weeks. This may be because the young need to be born with fat deposits so that they can go several days without food when their parents are away at sea.

The timing of incubation is usually fixed, so that the young hatch out within a few hours of one another. This means that their fledging date will also be approximately the same. Exceptions include the larger birds of prey, such as **eagles**, which lay two eggs, but incubate the first one immediately, so that it hatches a day or two before the second.

Fulmars have one of the longest of all incubation periods – up to eight weeks. This is mainly because the young bird needs to build up fat reserves to survive a long winter at sea.

Arctic Terns frequently mob anyone who approaches too close to their nest, and can give a nasty peck if you don't protect yourself against their attack.

Barn Owls, as their name suggests, are highly dependent on human habitations for their nests. They will also happily take to artificial nestboxes. The young huddle together for warmth and security.

PARENTAL CARE AND FLEDGING

Once the young have hatched, the hard work really begins. Young birds can be divided into two very distinct groups: those that are born blind and remain in the nest for several weeks, requiring constant care and feeding from their parents; and those that are able to leave the nest straight away, walk or swim, and find food for themselves or with the help of their parents.

The first group, which includes all the passerine birds (or **songbirds**) are known as altricial (or nidicolous) species; while the second group, which includes **wildfowl**, **waders** and **game-birds**, are known as precocial (or nidifugous) species. Some groups of birds, such as **gulls** and **terns**, are known as semi-precocial, as they are born able to see and with a downy coat, but stay in or very near the nest and are fed by their parents for some days after hatching.

Interestingly, both strategies work very well. Indeed, for them to be adopted by such a wide range of different species, they must both be evolutionary advantageous. For **songbirds**, which tend to hide their nest away in foliage or holes in trees, the altricial strategy means that their young are kept safe from predators, while they collect food in the immediate area of the nest. For aquatic birds, such as **ducks** and **swans**, and other precocial species, such as **game-birds**, all of which nest on the ground or on water, the strategy works equally well, enabling the young to range far and wide in search of food while avoiding predators by swimming, running or hiding. Fledging is a term that technically only applies to altricial birds, and refers to the period when the baby birds have acquired their first feathers (juvenile plumage), and can leave the nest and fly. Even at this time most baby songbirds are dependent on their parents for food and safety, and despite the adults' best attentions, this is the time when the death rate is the highest.

Fledging is also used to describe a similar process in precocial species, although it is not at the point at which they leave the nest (which is usually immediately or a few hours after hatching), but when they acquire their first feathers and are able to fly.

Like incubation periods, the time taken to fledge varies enormously from species to species and group to group. Most young songbirds fledge roughly 2–3 weeks after hatching. However, this period can range from just 11 days for the **Lesser Whitethroat**, and 12 days for the **Dunnock**, to 24 days for the **Swallow**. Crows stay in the nest even longer. Young **Jackdaws** and **Carrion Crows** take up to five weeks to leave.

For larger birds the fledging period can be much longer. Once again, seabirds hold the record, with young **Manx Shearwaters** taking up to ten-and-a-half weeks. The young ultimately must leave the burrow or starve, as the parents generally do not feed it for the last ten days before fledging.

HYBRIDIZATION AND UNUSUAL BREEDING BEHAVIOUR

HYBRIDS
One of the most puzzling things that the inexperienced birder might face is coming across a bird that, although it doesn't match any species in the field guide, appears to have characteristics of more than one

The Ruddy Duck, a feral species introduced from North America, has been accused of threatening its close relative, the White-headed Duck through hybridization.

species. Sometimes these are birds with plumage aberrations, such as partial albinos or melanistic birds. In other cases, they may be hybrids – the offspring of two different species.

The extent of hybridization varies dramatically. In some families, including many **songbirds**, it is virtually unknown. In others, notably **wildfowl**, it is very common indeed. For

Hybrids may be puzzling to the novice birdwatcher – like this cross between a Tufted Duck and a Pochard.

example, there are many recorded instances of hybrid **ducks** and **geese**, some of which have caused confusion among even very experienced birders, as they tend to resemble a third species rather than either of their parents (as in **Tufted Duck** x **Pochard** hybrids, which look like a rare American vagrant, the **Lesser Scaup**). **Geese** are particularly prone to hybridization, both between two wild species, and between one wild and one domestic bird. Birds of prey rarely hybridize in the wild, but misguided falconers have been known to artificially pair species, including such bizarre couplings as a **Merlin** x **Peregrine**.

Hybridization may appear to be a futile gesture, as the offspring tend to be infertile (indeed, the definition of a 'proper' species used to be that if it hybridizes with a close relative the young are not fertile). However, in a few cases, the two parent species share enough genes to result in a stable hybrid population being formed, which may ultimately evolve into a new race or species.

Perhaps the best-known recent case of hybridization is the **Ruddy Duck**, in which males of this introduced North American species have been dispersing from their British stronghold to southern Spain. Here they have mated with the rare **White-headed Duck**, diluting the gene pool and potentially threatening the entire south-west European population of this now endangered species.

ABERRANT BREEDING BEHAVIOURS

Another unusual type of breeding behaviour is known as reverse sexual dimorphism. This is extremely rare, but in Britain it is practised by two scarce breeding waders, the **Red-necked Phalarope** and **Dotterel**. In these species, the sexual roles are reversed to such an extent that it is the female who sports the brighter plumage, takes the lead in courtship and, having laid her clutch of eggs, leaves all the duties of incubation and care of the young to the male. In some cases female Dotterels have been recorded mating first in Scotland, then leaving the male to incubate the eggs while they fly to Scandinavia to mate once more and raise a second brood.

Perhaps the best known form of 'aberrant' breeding behaviour is the practice of laying eggs in other birds nests, known as brood parasitism. This is found in a number of the world's bird families, but only one species does it in Britain: the infamous **Cuckoo**.

Cuckoos arrive back from their African winter quarters in May, and immediately set about mating and laying their eggs. However, unlike more conventional species,

Cuckoos are well-known for their incredible habit of laying their eggs in other birds' nests. First, the female removes the host's own eggs (above), then the unsuspecting hosts raise the young Cuckoo as their own (left).

the female Cuckoo does not build a nest herself. Instead, she lays her egg in that of a host species (almost always the same species in whose nest she was herself raised). Typical hosts in Britain include the **Reed Warbler, Meadow Pipit** and **Dunnock**. First, she removes any eggs already laid by the host, before depositing her (usually camouflaged) egg with great speed. Once hatched, the young Cuckoo instinctively ejects any remaining eggs and/or chicks, thereby gaining a monopoly of their parents' attention.

The Cuckoo and its hosts are engaged in a constant form of escalating warfare, which the Cuckoo appears to be winning: by this method the female can lay up to 20 eggs, and still leave for Africa before her young have fledged.

POLYGAMY

Strictly speaking, polygamy should not be under the heading of 'unusual breeding behaviour', as it is practised by a large number of species. It used to be thought that while many birds (such as **Mute Swans**) pair more or less for life, in other species the male would have a harem of females, each laying his eggs in a different nest. A famous example of this practice (known as

Corn Buntings may look like unassuming little brown birds, but they have an extraordinary sex-life, with one male mating with up to seven females.

The Dunnock, too, has an amazing variety of breeding habits, with the male watching the female obsessively to avoid becoming the parent to another male's young.

polygyny) is the **Corn Bunting**, males of which may have seven or more females nesting within their territory. The **Cuckoo**, on the other hand, is polyandrous, which means that the female mates with more than one male.

However, scientists have discovered that many birds are not as faithful to their mates as was once thought. For example, the male **Dunnock** jealously guards his spouse, who otherwise will often mate with other single males in the area. Male Dunnocks have even resorted to tugging the previous male's sperm out of the female's cloaca, before re-mating to ensure that it is their offspring that is hatched. In other cases, females will take a partner to share nest building and incubation duties, but will still try to mate with other males. In this way, she maximizes her chances of producing the healthiest and fittest young, and passes on her genes to future generations. Males, however, must guard their mate to ensure that they do not become cuckolds to someone else's young.

MIGRATION AND NAVIGATION

In autumn and winter, flocks of Starlings, from northern and eastern Europe, move westwards to swell the British and northern European populations. At dusk, they gather in their thousands at a communal roost.

Swallows usually arrive in northern Europe from Africa in late April and early May. These two are taking time out from feeding to preen – essential to keep their feathers in good order.

WHY DO BIRDS MIGRATE?

It has been estimated that as many as five *billion* birds, of more than 200 different species, undertake the twice-yearly journey between Eurasia and Africa. These include such diverse groups as **waders** and **warblers**, **swallows** and **terns**, proving beyond doubt that as an evolutionary strategy, migration is a success.

Why do birds migrate? Why leave a comfortable home in Europe, and head thousands of miles south, facing hazards along the way? And if a bird manages to survive this perilous journey, and reach its winter home, what impels it to come back the following spring? Why not just stay put the whole year round? The traditional reason given is that these birds are unable to survive the northern winter and must head south to find food. This is true, but it is only half the story.

In fact, to view migratory birds as 'ours' heading south for the winter is to put the cart before the horse. Species such as **swallows**, **warblers** and **flycatchers** originated in Africa, and first headed north to avoid competition with other species. By travelling northwards, they found 'spare' ecological niches where they could breed and raise their young, with the advantages of abundant food, long daylight hours, and fewer competitors. To survive, they developed adaptations including longer wings, navigational devices, and the capacity to store large amounts of fat for the migratory journey. At the start of

After breeding, Swallows undertake an extraordinary journey halfway across the globe from Europe to southern Africa – a distance of almost 10,000 kilometres (6,200 miles).

autumn, however, as the temperature drops, the insects on which so many species rely for food begin to disappear. Meanwhile, the days get shorter, so there is less time available to forage for resources. For many insect-eaters such as warblers and flycatchers, there is only one choice to make: either stay put and starve to death, or head south for the winter.

We often assume that migrants face far greater dangers than resident species, but in fact the opposite may be true. Migration expert Peter Berthold has pointed out that while summer migrants, such as the **Sedge Warbler**, tend to produce a single brood of young, resident species such as the **Wren** or **Blackbird** raise two or more broods. Some species may lay more than a dozen eggs at a time. This would seem to suggest that staying put during the northern winter leads to a higher mortality rate than the journey back and forth to Africa.

The Swedish scientist Thomas Alerstam has gone even further, turning the question on its head. He asks why, given the obvious advantages of migration, any birds pursue the resident strategy at all – 'why do not *all* birds migrate?'

Sedge Warblers are summer visitors, producing a single brood of chicks before returning to warmer climes in the autumn.

49

HOW DO BIRDS NAVIGATE?

Ever since the Ancient Greeks tied messages to the legs of migrating birds, in the hope of discovering where they spent the winter, man has been fascinated by bird migration. Even now, when you look at a **Willow Warbler** or **House Martin**, it is hard to believe that these tiny birds are capable of travelling such great distances across the globe.

Indeed, until relatively recently, the prevailing theory suggested that, like many mammals, birds spent the winter months in hibernation. Sightings of **Swallows** gathering over water in autumn reinforced this theory, with observers claiming to see the birds plunge beneath the surface of the water. Even the great 18th-century ornithologist Gilbert White subscribed to this theory, at least in part, though he did have his doubts.

Yet, as long ago as the days of the *Old Testament*, people suspected the truth – that birds travel south in autumn to avoid the colder months. The best evidence for this early belief in migration can be found in the *Book of Jeremiah*: ·

'Yea, the stork in the heavens knoweth her appointed times; and the turtle-dove and the swallow and the crane observe the time of their coming.'

The concept that birds 'know the time of their coming' is surprisingly close to the truth. Many birds schedule their journeys according to changes in day-length, though local factors, such as weather conditions, also dictate timing of departure. Day-length affects the birds' endocrine system, producing hormones that stimulate them to prepare for the long journey ahead, for example by putting on extra fat supplies. Once they have returned, the same glands are responsible for prompting

Before migrating south for the winter, Swallows often gather in large flocks in reedbeds and on telegraph wires.

the onset of the breeding cycle. Once they have set off, birds use a complex hierarchy of cues to enable them to find their way, rather than relying on a single navigational method. These include navigating by the stars and sun, as sailors used to do, 'reading' the vibrations of polarized light over land and sea, and logging into the Earth's magnetic field as if they had an internal compass. They may also inherit route information as part of the genetic code passed on by their parents. Once migrants approach their destination, they rely more on visual cues such as local topography, often following coastlines or rivers.

None of these methods is infallible. Some birds, particularly juveniles undertaking their first migratory journey, appear to have something wrong with their internal 'clock', and may head in entirely the wrong direction. Others are diverted off-course by unusual weather conditions, especially cloud and rain, which make it impossible to use visual cues such as the stars or topography. Some groups of birds, especially migrants moving over large areas of water, appear to drift off-course with crosswinds, though they are usually able to reorientate themselves successfully.

House Martins also gather on wires, often calling to each other as they prepare for the epic journey ahead.

MIGRATION STRATEGIES

Not all birds migrate in the same way, or use similar strategies. Warblers and chats migrate by night, spending the day feeding or sheltering from predators. Others, such as swallows and raptors, travel by day, feeding as they go. The **Knot** and **Sedge Warbler** choose a 'long-haul' strategy, putting on huge amounts of fat before completing their journey in a few huge leaps of hundreds or even thousands of kilometres at a time. The **Swallow**, by contrast, potters from place to place and takes several weeks or even months to complete the journey to its final destination.

Some, such as the **North American warblers**, choose to fly across vast areas of sea in order to shorten their journey. Others, especially large birds of prey, such as **eagles** and **buzzards**, try to avoid crossing water at all, instead gathering in huge numbers at land-crossings such as at Gibraltar and around the Bosphorus in Turkey.

Each of these strategies has its advantages and disadvantages, but each has evolved over many generations to suit the particular species. It is only in recent years, with ringing studies and the ability to radio track larger species, such as the **Osprey** and **White Stork**, that we have begun to understand the subtleties of migration strategy.

Not all species use the same strategy for both the outward (autumn) and return (spring) migrations. For example, those North American warblers that head over the open ocean in

Montagu's Harriers undertake a long journey south each autumn from their western European breeding grounds to spend the winter in West Africa. Their long wings and buoyant flight action enable them to cover long distances without huge effort.

autumn follow a quite different route in spring, when they head along the coast in short hops. This is because in spring there are no suitable tail-winds to enable them to cross the

ocean in the two or three days needed to reach landfall. This strategy, known as 'loop migration', is also followed by several European species, including the **Red-backed Shrike**. In autumn, these birds take a south-easterly route out of Europe, via the eastern Mediterranean to winter quarters in Central and Southern Africa. In spring, they return by an even more easterly route, across the Arabian peninsula, probably because meteorological factors make this course more favourable. British **Sand Martins**, too, follow an anti-clockwise route, heading south via Iberia and north-west Africa, but returning via the

Sand Martins are long-distance migrants, with birds spending the winter well south of the Sahara Desert in Africa.

Spoonbills are short-distance migrants, with most European birds wintering in the Mediterranean basin or North Africa.

53

Common Cranes are amongst Europe's best-known migrants, as they are seen passing overhead in vast flocks, calling to each other with loud, honking cries. Cranes breeding in Scandinavia may winter in France or as far south as Spain.

central Sahara, Italy and central Europe. Even within a particular species, different populations pursue different strategies. One of the most amazing of these is known as 'leapfrog migration'. This involves a population that is breeding at a higher latitude 'leapfrogging' over a population of the same species breeding at a lower latitude. So, for example, Arctic populations of wader, such as the **Ringed Plover**, migrate all the way to sub-Saharan Africa, while birds breeding in southern Scandinavia travel only as far as southern Europe or North Africa. British breeding Ringed Plovers follow an even more sedentary strategy, being more-or-less resident, although some do move westwards to areas of milder winter weather.

Ecologist Thomas Alerstam of Lund University, Sweden, has proposed two possible explanations for the existence of leapfrog migration, both to do with competition and timing. The first theory suggests that more southerly birds finish breeding earlier, and move to the nearest available suitable wintering areas. More northerly birds finish breeding later, and by the time they travel south, these areas are already 'full', being occupied by the southerly breeders. This forces the northern populations to head farther south still, until they come across somewhere to spend the winter.

The second, more plausible theory, is to do with the timing of breeding itself. Alerstam suggests that it is crucial for birds breeding in temperate zones to winter near their breeding grounds, as the start of their breeding season, being primarily determined by the local weather conditions, may vary considerably.

In contrast, Arctic breeders have only a narrow window of opportunity during which conditions are suitable to breed. Therefore, they must return to their breeding grounds at more or less the same time every year, whatever the weather conditions. They rely on a sophisticated internal timing mechanism to tell them when to return, so have no need to spend the winter close to their breeding areas.

Ringed Plovers have a complex migration strategy, with some populations heading much farther north or south than others, which remain on or near their breeding grounds the whole year round.

UNUSUAL MIGRATION

Not all migratory journeys go to plan. Indeed, one of the most fascinating areas of birding is observing what happens when things go wrong on a migrant's journey.

One of the more common migration-related phenomena is that of 'falls' – a sudden and unexpected arrival of migrating birds, usually, though not always, at a coastal 'hotspot'. Falls are generally weather related, as it usually takes some sort of

In spring, British birders hope for warm, fine weather from the south, which may bring overshooting migrants such as this Hoopoe.

TOP TIP

Weather and migration
Watch the weather forecast for unusual weather events, such as autumn gales or heavy rain. These may bring falls of songbird migrants, or 'wrecks' of seabirds, which sometimes appear far inland, on reservoirs or gravel-pits. Despite being blown far off-course, they usually manage to reorient themselves and get back to sea.

adverse weather conditions to concentrate the birds and force them to land. For example, during the autumn, migrant songbirds usually depart from Scandinavia during anticylonic weather, which provides clear skies and light following winds to help them on their way. If conditions remain good, they will cross the North Sea and make their way along the coasts of continental Europe, towards Africa.

But when low-pressure systems are present over the North Sea, migrants often get disoriented, and blown off-course by the strong winds. Many drop exhausted into the sea and die. But others carry on and, with luck, make landfall somewhere along the coasts of eastern Britain.

Most **songbirds** migrate by night, so falls often occur in the early hours of the morning. Some can be spectacular, involving many thousands of birds of many different species. But if you want to experience the thrill of seeing a fall, you have to be quick. Once the birds have rested, fed, and recovered their strength, they are off again, impelled by that mysterious, migratory urge.

Another phenomenon occurs in spring, and is known as overshooting This involves those species that return each year to breed around the shores of the Mediterranean Sea or in France, such as various species of **heron** and **egret**, **Red-rumped Swallow**, **Woodchat Shrike**, **Alpine Swift**, **Hoopoe** and **Serin**.

The various European species of crossbill are dispersive, irruptive migrants, heading to different areas each year in order to breed.

The key factor in overshooting is the weather, not just in Britain, but farther south, over Europe. Ideally, there should be a large area of high pressure, or anti-cyclone, situated over the northern shores of the Mediterranean, with its northern limits reaching southern Britain. As the birds arrive at or near their normal breeding-grounds, the clear skies and light, southerly winds allow the birds to continue past their destination.

Apart from the meteorological factors, why do birds overshoot at all? One possible explanation is that a tendency to overshoot is genetically programmed, it being a potential advantage for the bird concerned to discover somewhere new to breed. Back in spring 1987, a pair of overshooting **Black-winged Stilts** raised two young at Holme in Norfolk, showing that, for

Waxwings are common breeding birds in the far north of Europe, which undertake irregular, irruptive migrations to the south and west, depending on the availability of food.

some birds at least, the pioneering spirit can lead to success. In a sense, most vagrant birds are 'lost' birds. They will usually perish, but just occasionally, form the pioneering contingent of a new colony. In recent years former vagrants such as **Cetti's Warbler**, **Mediterranean Gull** and **Little Egret** have all become regular British breeding species.

Another form of pioneering spirit can be seen in those few species that are 'irruptive', that is, species that occasionally leave their normal breeding and wintering areas to arrive elsewhere in huge numbers. Of these the commonest and most familiar in Britain is the **Crossbill**, which regularly turns up in mid-summer, and may arrive in large numbers in areas previously devoid of the species. Crossbills generally stay put until the following year, when, after breeding early in the New Year, they head off again on their nomadic journeys. Another irruptive species, the **Waxwing**, is an autumn and winter visitor here. In some years thousands may arrive, in others hardly any. The size of Waxwing irruptions is governed by the lack of berry food on their northern breeding grounds.

Two other species, the **Nutcracker** and **Pallas's Sandgrouse**, are much more occasional irruptive visitors, though when they do occur it can be in huge numbers, to the delight of twitchers.

Not all birds migrate huge distances. An often overlooked phenomenon is known as 'altitudinal migration'. As its name suggests, this involves a journey from higher to lower altitudes, usually to spend the winter months where there are more accessible food supplies. Although these journeys may seem insignificant in terms of distance travelled, they often involve

The Ptarmigan undertakes short movements up and down the mountains where it breeds, depending on food and the local weather conditions.

a major change of lifestyle. In winter, **Skylarks** breeding on Britain's upland moors and mountains head to the lowlands, often gathering in flocks near the coast. Even true 'high-altitude' species, such as **Ptarmigan**, will move locally down a mountain, especially if heavy snow covers their food supplies.

One species makes a doubly unusual migratory journey, moving both downwards to lower altitude, yet also northwards for the winter. The **Water Pipit** breeds in the high Alps and Pyrenees, yet a significant population spends the winter in southern Britain, usually near fresh water where the birds can feed on insects.

For another unusual migratory strategy, we must turn to three species of seabird, the only ones to reverse the prevailing north-south migratory trend. **Wilson's Storm-petrel**, **Great** and **Sooty Shearwaters** all breed in the southern hemisphere, then head north across the Equator to spend their 'winter' (our summer) in the northern hemisphere, before returning south during our autumn in time to breed once again.

DISTRIBUTION AND RANGE

Golden Eagles are the classic bird of the high mountains of Scotland, where they feed on a variety of prey, including grouse, Ptarmigan and mountain hares.

TOP TIP

Edge habitats
The point at which two different habitats meet, called an edge habitat, is often very productive for birds. Check out the borders between woodland and farmland, heath and scrub, or salt and freshwater marshes. Hedgerows are the ultimate 'edge' habitat, and are often full of nesting birds.

HABITATS: THEIR INFLUENCE ON BEHAVIOUR

At first sight, a bird's habitat may not seem to have much direct influence on its behaviour, but a closer look reveals all sorts of subtle effects. For example, wader species that feed mainly on coastal marshes and estuaries are highly influenced by the twice-daily movements of the tide. The need to feed at low tide and roost at high tide influences their entire diurnal rhythms, which must be changed accordingly. The feeding methods of **birds of prey** are also influenced by their habitat, and vice versa. In areas where they mainly prey on light items such as Ptarmigan, **Golden Eagles** may nest fairly high up a mountainside. In the west of Scotland, however, where they feed mainly on heavy prey such as rabbits, they need to build their nest at low altitudes in order to carry their prey down the hillside rather than up, which would use more energy. **Sparrowhawks** have evolved a compact shape with rounded wings and long tail in order to manoeuvre themselves through dense foliage, while falcons, such as the **Kestrel** and **Hobby,** which hunt aerially, are more streamlined in shape.

Bird song, too, is influenced by habitat. Indeed, it might even be argued that the different varieties of song have evolved to suit different habitats. Woodland species such as the **Blackbird, Blackcap** and **Nightingale** tend to have rich, fruity songs, in order to penetrate the foliage around them to the greatest effect. With so many surfaces to absorb sound, a complex song is the only way to ensure that the message is properly heard. Species that live in marshes and reedbeds adopt a different strategy. Many of their calls and songs, from species as diverse as **warblers** and **crakes**, are monotonous, rhythmic and repetitive. Indeed,

The Nightingale's fluty song has evolved to penetrate the dense foliage of its woodland habitat. It is thought that by singing at night it avoids competition with other songsters, so that it can be heard more effectively.

Water Rails have a bizarre screeching call, which sounds to many people like a pig in distress.

The prize for the most bizarre noise made by any British bird might go to the Grasshopper Warbler, whose insect-like song sounds a bit like a fishing reel being played out.

they are often more similar in sound to marsh-dwelling species of amphibian or insect than to other songbirds. Many years ago, the natural history sound recordist John Burton compared the songs of **Savi's Warbler** and **Grasshopper Warbler** with the sounds of amphibians and insects, such as the Marsh Frog and the Wart-biter Bush Cricket. He concluded that the songs and calls had evolved in parallel to suit the acoustic nature of the habitat. It seems that monotonous sounds are more effective in monotonous habitats.

Finally, where birds choose to nest influences their behaviour and appearance to a large degree, especially regarding whether birds show sexual dimorphism, that is, different plumages for males and females. For example, hole-nesting birds such as the **tits** and **Tree Sparrow** can afford to have a brightly coloured plumage even in females, as they are not vulnerable to predators on the nest, while birds that nest in more open areas such as **larks** and **pipits** tend to be brown and streaky in colour for effective camouflage. Vulnerable species, such as **ducks**, also show sexual dimorphism, with bright, colourful males taking the attention away from the duller, cryptically marked females. Large, aggressive species such as **Mute Swans** however, can afford to be showy and noticeable.

RANGE AND DISTRIBUTION

The great ornithologist James Fisher once wrote that a bird does not have a range, only a 'current range'. He was referring to the fact that during as short a period as a single human generation – as little as two or three decades – the range of a particular species can alter drama-tically, either expanding to colonize new areas, or con-tracting and disappearing from former haunts.

Even since the late 1960s, during the 30 years or so that I have been watching birds, I have witnessed a number of dramatic changes in the range and status of our breeding, migratory and wintering birds. The decline of farmland species such as the **Skylark**, **Yellowhammer** and **Grey Partridge**, for example, are caused almost purely by outside influences such as modern farming methods, and are therefore outside the scope of this book.

Another, more positive, change has been the welcome increase in the population and range of birds of prey, such as the **Buzzard** and **Hobby**, this time due to the banning of harmful

The Skylark once had the widest distribution of any British bird, but in recent years, its range and numbers have contracted due to unsympathetic modern farming methods, such as wheat and barley monoculture.

The Hobby's population has increased faster than any other breeding raptor in the past few decades, and the bird is now a regular sight across much of southern Britain.

pesticides such as DDT. But other changes in range are more complex, and may be influenced either wholly or in part by the behaviour of a particular population of birds.

Perhaps the best-known examples of dramatic changes in range are where populations have suddenly expanded, such as the **Little Egret** and **Mediterranean Gull**, which have colonized Britain as breeding species in the past decade or so. Again, human influence may be a factor, with climate change giving a helping hand to both these species. But other factors may also be at work, notably some kind of genetic mutation that has allowed individual birds of each species to act as pioneers, followed by an eventual full-scale colonization.

Another example involves a common British breeding species that now also winters here in good numbers. Wintering **Blackcaps** began to be reported

One of the newest arrivals as a British breeding bird is the Little Egret, which has spread northwards from its Mediterranean stronghold.

In recent years, Blackcaps have begun to spend the winter in Britain, where they survive, thanks to mild weather and a plentiful supply of food, such as berries, in addition to peanuts and seeds provided by humans.

in various parts of the country two or three decades ago. At first, most people assumed these were 'our' breeding birds that had decided to stay put for the winter. Studies then revealed that they were actually German Blackcaps. Instead of migrating south-west to spend the winter in Spain, Portugal or North Africa, they had headed north-west and ended up in Britain.

Two factors enabled these birds to survive: first, mild winter weather prevented mass death by starvation; second, the German birds adapted their feeding behaviour in order to take advantage of the plentiful seeds and other foods provided by humans. As a result, they survived and thrived, and returned to Germany to breed ahead of their rivals. Twenty years later, and the entire population from this region now spends the winter in Britain.

As climate change adds yet another factor in the myriad influences on the range and distribution of Britain's birds, birders can look forward to many changes during the next few decades – some for the better, others for the worse.

LIFE AND DEATH

Juvenile Starlings have a dull brown plumage, but moult in late summer into the familiar spotted garb of their parents.

This may seem to be rather a forbidding section: the very title may put some readers off. It is certainly true that the subjects covered do not all lend themselves to careful field observation. Nevertheless, all these are part of a bird's lifecycle, and therefore vital to an overall understanding of bird behaviour in its broadest sense. I have tried to avoid too many technical terms (though some are inevitable), and, where possible, to give examples when a particular type of behaviour can be observed.

MOULTING AND PLUMAGE

Birds do not keep all their feathers for the whole of their lifetime. Indeed, most undergo an annual moult, during which all or part of their plumage is replaced by spanking new feathers, giving the bird a better chance of survival.

Moult occurs for several reasons. The first, and most important, is to enable birds to fly as well as possible. Old, broken or worn feathers reduce efficiency, and can lead to the bird being less able to find food, or more vulnerable to attack by a predator. Therefore, having the best possible quality of plumage is absolutely vital. The second reason is that old, worn feathers are also less efficient as insulators, and, in cold, winter weather, the quality of a bird's plumage can make a major difference between life or death. Finally, old and worn feathers tend to fade in colour and brightness, and, for many birds, the quality of plumage – in particular, specific colours and markings – is what helps them to attract the best possible mate. For all these reasons, an annual moult is pretty essential. But when should it happen? The problem with even a partial moult is that during the intermediate stage – when some feathers are old and others new – the bird is at its least effective in terms of flight, keeping warm or attracting a mate. For this reason, almost all birds undergo their major moult after the end of the breeding season, but before the onset of autumn and winter when food resources are more scarce and the temperatures are lower.

There are other advantages to moulting in late summer, especially for small **songbirds**. Feeding a brood of hungry young leaves plumage in an exceedingly tatty state. At this time of the year, too, there is plenty of foliage in which to hide to avoid predators, and there is a plentiful and freely available supply of food. So don't be surprised when, after months of songbird activity in your garden or local woodland, everything suddenly goes quiet and the birds appear to have vanished. They are probably still around, but just lying low. Another group of birds that moults from mid to late summer is the **ducks**. Many species

TOP TIP

Moulting
When birds are moulting, they are often hard to identify on appearance alone, as they may look nothing like they do when in full breeding or non-breeding plumage. So concentrate on both jizz and behaviour to establish a bird's identity. For example, feeding behaviour is often a good guide.

This Mallard drake in 'eclipse' plumage looks very different from his breeding garb. Mallards moult during the summer months, from June to August.

of duck go into a stage known as 'eclipse' plumage, during which males adopt a very dull plumage superficially similar to the female's, though usually with a few clues as to the bird's true identity. Male **Mallards**, for example, lose their glossy green head pattern, while male **Tufted Ducks** go from smart black and white to a chocolate brown very similar to that of the female. This often confuses inexperienced birders, who may think that all the males have vanished. During the eclipse period, the drakes may even lose the power of flight, and become much more wary than usual. Then, from roughly early August onwards, the various species reappear in their fine new feathers, as if by magic. Because they are vulnerable to predators at this stage of their lifecycle, some ducks, including the **Shelduck**, gather in huge flocks for safety against predators.

Other birds are unable to enjoy the luxury of flightlessness, as they must constantly catch prey. They therefore moult gradually. Day-flying birds of prey, such as **hawks** and **buzzards**, follow this pattern, gradually shedding and regrowing their flight feathers in systematic order, so enabling them to hunt. For this reason some raptors may be in a virtually permanent state of moult. Young songbirds hatched in the spring also undergo a period of moult –

In his spotted plumage, this juvenile Robin is very different from his red-breasted parents. It will retain this juvenile plumage until early autumn.

Young Black-headed Gulls lack the dove-grey upper parts and brown hood of their parents.

initially, from the downy covering into their first true plumage, known as 'juvenile', then, usually a few weeks later, into the full adult garb. Baby **Robins** pass from downy fluff into the spotted browns and buffs of juvenile plumage, finally emerging, around two months after fledging, into their glorious adult plumage with the orange-red breast.

For long-distance migrants, such as the **Whitethroat**, the moult is a vital prelude to getting themselves in the best possible condition to travel thousands of kilometres south after breeding.

Some species undergo more than one moult a year, into and out of what used to be called 'summer' and 'winter' plumages, but are now usually referred to as 'breeding' and 'non-breeding' plumages. Species that follow this pattern include divers, **grebes**, **gulls**, and **terns**. As well as these two stages, gulls also go through a number of other plumages during the three or four years it takes them to reach full maturity. These moults can lead to confusion among birders, as the differences between plumages can be quite subtle, and not all birds follow exactly the same moult stages.

BATHING, PREENING AND FEATHER CARE

Closely related to moult is the whole business of caring for feathers, involving a range of behaviours such as bathing (in dust and/or water), preening, and general feather maintenance.

Birds must bathe regularly in order to keep their plumage in tip-top condition, and free from dirt and parasites. They do so in a number of ways, of which the most commonly observed is simply bathing in shallow water. This is usually accompanied by a complex and choreographed set of movements designed to cover the whole of the bird's plumage with water in the most efficient and effective way possible.

Virtually all species of bird bathe, but a small minority do so not in water, but in dust. At first this seems bizarre. After all, how can covering your feathers with dust keep them clean? Yet it does appear to work, apparently by removing oil, grease and parasites from the feathers. **House Sparrows** seem to be particularly fond of 'dusting'.

Some birds, especially small **songbirds**, tend to bathe in a shallow puddle or edge of a pond or stream; others, including **waders**, may do so in deeper water. True waterbirds, such as **ducks**, **grebes** and **coots**, will bathe while swimming, submerging

Bathing is essential for birds, to keep their feathers clean and neat; this House Sparrow is enjoying the water in a bird bath.

part of their body under the surface and letting the water wash over them until they are clean.

After bathing, especially on a fine, warm day in spring or summer, birds often spend time sunbathing. This enables them to dry off their feathers and heat themselves up. **Blackbirds** are particularly fond of this. Other birds, notably **Cormorants**, hang out their wings to dry in the sun and wind – something they have to do, because unlike ducks and other waterbirds, their feathers are not waterproof.

Following a bath, birds also spend time putting their feathers in tip-top condition by preening, usually with their beaks, but occasionally involving scratching with their feet as well. Preening is vital: it puts the feathers back where they should be, straightens out any problems, and allows the bird to remove any dirt or parasites that may have survived the bathing process. Some species, especially **waterbirds**, secrete oil from a gland near the beak, which they use to waterproof their plumage.

Cormorants do not have the ability to 'waterproof' their plumage, so must hold out their wings to dry.

SIGHT, HEARING AND SMELL

Of all the senses, birds rely most on sight – a critical factor in the ability to find food, avoid predators, and during courtship rituals. One of their most developed abilities is that of 'visual acuity' – or being able to distinguish between tiny differences – say between small particles of dirt and those of food. Even day-old chicks can tell the difference between very similar objects, and differentiate between subtle shades of colour.

Indeed, birds' colour vision is superb. This is another vital requirement, especially when searching for a mate, where tiny differences in the colour and brightness of a male's plumage enable the female to judge his state of health, and therefore whether he will make the best father to her young. One way in

Like other members of the owl family, Barn Owls have superb night vision, which enables them to catch their prey by hunting on the wing.

which birds' eyesight differs greatly from that of humans is the position of their eyes. Most birds have eyes on either side of their face, which gives them excellent all-round vision for detecting food or predators, but means that they lack the binocular vision that we take for granted, and that helps us to judge perspective. The obvious exceptions to this rule are **owls**, which have both eyes facing forward, enabling them to spot and catch their prey more easily.

Another way in which birds differ from ourselves is their ability to see ultraviolet light, enabling them to appreciate colours and shades that we are unable to see. This is especially useful for birds that feed on fruit and flowers, such as **hummingbirds**. The ability to detect ultraviolet light may also help migrating birds, especially when cloud cover obscures the sun.

Hearing is another vital sense for many birds, especially predators such as **owls**, which may hunt almost entirely using sound cues, especially if their prey is under snow or dense foliage. The ability to detect distant calls and songs is also vital, either to enable birds to hear warning calls as a predator approaches, or for females to hear distant males. Birds also have an extraordinary ability to differentiate calls, this is especially useful for colonial nesting seabirds such as **gulls**, **petrels** and **shearwaters**, where the returning adult finds the nest by listening for the young birds' calls.

Finally, smell is also used by particular groups of birds to help them find food. Seabirds have an especially well-developed sense of smell, a vital ability when food resources may be distant from each other at sea.

European Storm-petrels – the UK's smallest breeding seabird – have an extraordinary sense of smell and are able to find food at sea from vast distances away.

Seabird colonies, such as this huge breeding group of Guillemots, are not the most hygienic places, as the birds excrete where they stand.

EXCRETING WASTE

Just like any other living creature, birds must get rid of waste products that would otherwise build up in their bodies and cause harm. The fact that by doing so they cover our car windscreens is simply a matter of chance.

Unlike mammals, birds excrete their urine and faeces through the same opening, the cloaca, which is also involved in the reproductive process. Birds' waste products vary considerably, depending on diet. Seed-eating birds produce dry droppings; those that eat moist food with a high water content, such as fruit and insects, produce the liquid droppings that cause us more inconvenience. Indeed, some droppings can actually be dangerous to human health. Colonies of **feral pigeons** in particular may carry lung diseases such as psittacosis, which in some cases can prove fatal.

Another way in which birds excrete is through their glands. True **seabirds**, such as **shearwaters** and **petrels** have adapted ways of expelling excess salt from their drinking water and food, which otherwise might build up in their bodies and kill them. They do so by means of glands just above their bills, which get rid of up to 90 per cent of the salt in their diet.

TEMPERATURE REGULATION

Like all 'warm blooded' animals, including humans, birds regulate their body temperature by means of internal processes, rather than relying on external factors such as the sun, as 'cold blooded' animals such as reptiles and amphibians do. Birds manage their body temperature in response to changes in the outside environment by a process known as thermo-regulation. Nevertheless, they can suffer problems with both overheating and severe cold, and have adopted a range of different behavioural strategies to deal with these.

Small birds, such as **songbirds**, are especially vulnerable to rapid changes in temperature. Their larger surface-area-to-volume ratio means that their bodies lose or gain heat much more quickly than those of larger birds.

During spring or summer mornings, many species, such as the **Blackbird**, will warm up their bodies by sunbathing – spreading out their feathers in order to gain maximum benefit from the sun's rays. Later in the day, if the weather gets very hot, they must reduce their body temperature or risk overheating. They may do so by bathing, but because they do not have sweat glands, they cannot sweat away moisture to keep cool as humans do. Instead, they have to pant to allow moisture to escape from their throat

This Blue Tit is removing a faecal sac – a kind of 'shrink-wrapped poo' – from its nestbox to keep the nest hygienic and clean.

House Martins often huddle together for warmth, especially on cool days in spring, when they must avoid losing heat too rapidly, or die.

TOP TIP

Temperature regulation
Birds often vary their appearance dramatically, depending on the temperature of their surroundings. Bear in mind that a bird that has fluffed up its feathers against the cold may appear quite different from what you are used to, which could result in misidentification.

and breathing passage. They may also seek shade, which explains why small birds are often very hard to see during hot summer days.

Cold weather creates a very different problem: the rapid loss of heat from the unfeathered 'bare parts', such as the bill, legs and feet. To counter this, many small birds roost together, huddling up as close as possible in order to take advantage of their collective body warmth. During the short winter days, birds retain heat by fluffing out their feathers, trapping pockets of warm air beneath. This may give them a very different appearance from normal, and make identification difficult.

Waterbirds, such as **ducks**, **geese** and **swans**, generally find temperature regulation easier, as during very hot weather they can simply immerse themselves in cool water. In severe cold, they conserve heat by standing on one leg on the ice, which reduces heat loss through their feet.

BIRDS AND WEATHER

As the great American bird artist Roger Tory Peterson once said: 'Birds have wings – they travel.' By spending so much of their lives in the air, birds are surely influenced by the weather more than almost any other living creature. As a result, birds have

Winter thrushes, such as the Redwing, often arrive in large numbers from Iceland and Scandinavia when colder weather forces them to move farther south.

On warm summer evenings you may see Black-headed Gulls swooping overhead, feeding on flying ants.

gained a reputation as excellent weather forecasters. They often alter their behaviour as a result of changes in weather conditions, and, by observing this, our ancestors were able to predict the coming weather for themselves. Much of this knowledge has been passed down from generation to generation in rhymes and proverbs, which make up a unique body of weather folklore.

For example, the 'tumbling' behaviour of **Rooks** in autumn is supposed to foretell a change in the weather, probably because this behaviour tends to occur during windy conditions that usually signify the coming of a depression. Insect-eating birds, such as **swallows** and **martins**, also change their behaviour depending on the current weather. During settled spells of high pressure they feed on insects high in the sky, while during changeable periods of low pressure they tend to come lower, following their insect prey. By observing this behaviour it is possible to forecast the following day's weather with some accuracy.

Other species, such as **woodpeckers**, are often associated with rain. Woodpeckers have a habit of calling and drumming before the arrival of bad weather, and this has given them the name 'rain bird' in many parts of Britain, Europe and North America. In

TOP TIP

Behaviour and weather
Birds may behave very differently from usual during spells of extreme weather: notably prolonged drought, an extreme cold snap, or strong winds. So watch the weather forecast, and, during unusual weather conditions, look out for changes in bird activity on your local patch.

Shetland, the **Red-throated Diver's** habit of calling when rain is expected has earned it the folk name of 'rain goose'.

Birds are also affected by extremes in weather, with harsh winter weather perhaps the most serious threat. Small birds, such as the **Wren** and **Robin**, are particularly at risk, as they must eat around one quarter of their body weight every single day if they are going to survive. When the ground is covered with a layer of snow, or when freezing temperatures cause branches and twigs to ice over, then they simply cannot get to the seeds or insect food they require. As a result, many species change their behaviour, with normally shy birds such as **woodpeckers** and the **Nuthatch** becoming quite bold, often visiting bird tables to get access to a ready supply of food.

The weather can also affect birds during the breeding season. If they are early breeders, such as the **Blackbird**, a late cold snap may reduce their food supplies at a crucial time. Later on in the spring, cool, wet weather in May and June will reduce the chances of eggs hatching. Even if the eggs do hatch, the parents may not be able to obtain enough food to satisfy their hungry chicks. This is especially crucial for insect-eating species, such as tits and warblers, or birds on the northern edge of their range in Britain, such as the **Golden Oriole**.

The Mistle Thrush earned its country folk-name of 'Stormcock' because of its unusual habit of singing during heavy winds and rain. No one knows why it does this.

Every spring and autumn, migrating birds travel huge distances across the globe, as they seek out the very best places to breed and to spend the winter. Along the way, they encounter all kinds of weather, from helpful following winds to potentially fatal gales, storms and hurricanes. Many fail to survive the journey. Those that do need a mixture of instinct, good luck and an ability to deal with weather systems, that has evolved over many generations.

In spring, returning migrants often delay their arrival by as much as a week while they wait for good weather over the Channel to allow them to make the crossing safely. Easterly winds may bring drift migrants that have been blown off-course on the journey from mainland Europe to Scandinavia, across the North Sea to land on Britain's east coast. During late summer and autumn, small birds are at a greater risk from the weather. Migrants heading south from Scandinavia wait for a cold front which provides clear skies and following winds, ideal for crossing the North Sea. Sometimes the birds get things wrong, or as they cross the sea, may hit the poor weather that is associated with depressions. They become disoriented, and many perish, falling exhausted beneath the waves. Others fly on through the wind and rain, making landfall on Britain's east coast, to the delight of birders dedicated enough to venture out in the bad weather. Even more extraordinary is the annual arrival of North American passerines in Britain. These birds have been swept across the north Atlantic by strong westerly gales, to arrive exhausted in vagrancy hotspots such as the Isles of Scilly.

In the longer term, global climate change threatens to affect the lives of birds more than any other factor. We have already seen some species, such as the **Little Egret** and the **Mediterranean Gull**, shift their breeding ranges northwards as a result of climate change. Breeding species such as the **Bearded Tit**, **Hobby** and **Nightjar**, once confined to southern Britain, are now beginning to extend their ranges northwards. Meanwhile, at the other end of the country, three birds of the high tops, the **Snow Bunting**, **Dotterel** and **Ptarmigan**, are likely to disappear as British breeding birds, as a result of major habitat change.

Every cloud has a silver lining, and in this case it comes in the form of new species colonizing Britain from the south, and possibly also from the east. These may include such exotic creatures as the **Hoopoe** and **Bee-eater**, both of which now breed within reach

The Golden Oriole may expand its breeding range in Britain as a result of global climate change.

If temperatures continue to rise, European breeding species, such as the Bluethroat, may become regular breeders in Britain.

of Calais, as well as less glamorous birds such as the **Great Reed Warbler** and the **Black Kite**, which is one of the world's most adaptable and successful raptors.

If global warming brings a more continental climate, then several species with a more easterly distribution, such as **Common Rosefinch** and **River Warbler**, could find eastern Britain a suitable place to colonize. Whatever happens, the birders of Britain and Ireland can look forward to an exciting time in the next half century.

DISEASE AND DEATH

Like all creatures, birds suffer their fair share of disease. Indeed, along with killing by predators, and lack of food, disease is one of the three major causes of death in wild birds. Those especially at risk include young birds just out of the nest, whose immune system may not be quite as well developed as that of the adults, and older birds, whose bodies may have been weakened by the toll of raising successive broods of young.

Another problem is that of epidemic diseases, which can affect both colonial species such as **seabirds**, and sociable species such as **Starlings** and **House Sparrows**. Diseases, such as salmonella have increased in recent years due to the artificial concentration of birds in areas where humans feed them. This is a very good reason to keep your garden bird-feeding station clean and to discard old or mouldy food regularly.

Many bird species also carry unwelcome guests in the form of parasites. These may be either endoparasites, such as liver flukes or tapeworms, which live inside the bird's body, or ectoparasites,

Colonial breeding species, such as Gannets, are especially vulnerable to epidemics of disease, such as biological toxins, which may devastate a colony.

This Black-headed Gull is a victim of an unpleasant and often fatal epidemic disease known as botulism.

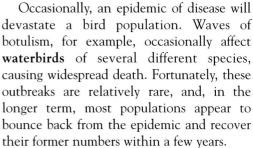

Birds that feed regularly at feeders and bird tables, such as Great Tits, may be vulnerable to diseases, such as Salmonella poisoning.

such as fleas, feather lice and mites, which live on the outside, usually underneath the feathers. Many of these parasites are unique to particular bird species and have co-evolved with them over many thousands of years. They may cause some harm, even leading to death in a few cases, but for most host species, they appear to be accepted as an occupational hazard.

Occasionally, an epidemic of disease will devastate a bird population. Waves of botulism, for example, occasionally affect **waterbirds** of several different species, causing widespread death. Fortunately, these outbreaks are relatively rare, and, in the longer term, most populations appear to bounce back from the epidemic and recover their former numbers within a few years.

As to the difficult question of how long birds live, generally, the larger the bird, the longer it is likely to live. A rough rule is that average longevity is correlated with body weight, using a complex mathematical formula. A species weighing approximately 32 times as much as another will live for around twice as long. So most songbirds such as the **Robin** and **Blue Tit** have a mean longevity of only one or two years, and a maximum lifespan, in a very few cases, of perhaps seven to ten years. There are, of course, a few individual exceptions to the rule of live fast, die young: ringing recoveries have included a **Blackbird** and **Starling** aged 20 years, a 16-year-old **Swallow** and a 15-year-old **Great Tit**.

Birds of prey, such as **eagles**, do not become fully mature for five or more years after hatching, and may live as long as 20 or even 30 years. Surprisingly, perhaps, the oldest recorded wild birds are somewhat smaller in size: an **Oystercatcher**, which survived for 36 years after being ringed (a good advertisement for a shellfish diet, perhaps?), and a **Fulmar**, which lived for an incredible 50 or more years.

Birds that generally live longer tend to follow a strategy of low clutch-sizes followed by a long period of parental care (as in the larger **seabirds**), as opposed to large clutches and early fledging (as in the case of most **songbirds**).

Finally, there are always exceptions to the longevity rule – when birds are kept in captivity. Safeguarded from predators, disease and other life-threatening factors, birds can live as long as their owners. The record goes to a **Sulphur-crested Cockatoo** in London Zoo that was known to have been over 80 years old when it died in 1982.

Waders tend to live a long time – one Oystercatcher survived for 36 years after being ringed as a chick.

2 SPECIES BEHAVIOUR

This section of the book deals with behaviour, organized family by family and species by species. There will be some overlap with Part 1, but, by grouping together related species, the book covers distinctive behaviour patterns, such as flocking by tits in winter. It examines both typical and unusual aspects of behaviour, and includes the 200 or so species you are most likely to encounter in Britain and northern Europe.

This second section of the book aims to:
- Provide an easy reference to a group or species.
- Give helpful hints to aid identification.
- Give a deeper insight into behaviour.

In some cases, similar families have been grouped together for ease of use, such as the section on seabirds, which includes unrelated species such as shearwaters and auks that share the same habitat.

DIVERS AND GREBES

Divers and grebes are both families of waterbirds, characterized by their ability to swim and dive. Unlike true wildfowl (ducks, geese and swans), they have only partially webbed feet, which are towards the rear of their body, making them well adapted to water but clumsy on land.

Two species of **diver**, the **Red-throated** and the **Black-throated**, breed in Britain, while the third, the **Great Northern Diver**, is a regular winter visitor. However, all three species are more likely to be encountered outside the breeding season, generally offshore, though occasionally on inland gravel-pits or reservoirs. They are generally identified by their distinctive jizz, caused by sitting very low in the water, often with only the head, neck and part of the back visible. As their name suggests, these birds dive frequently, and they are often submerged for long periods of time. They are highly mobile underwater, sometimes reappearing a considerable distance from where they originally dived. Despite their appearance, the divers are also strong fliers, usually flying low over the waves in a very direct manner. During the breeding season they nest at the edge of small lochs, where they may draw attention to their presence by their plaintive call. In Shetland, the Red-throated Diver is known as the 'rain goose' because of its supposed habit of calling when it is going to rain.

Do remember that nesting divers are easily disturbed, and that you should not approach them too closely.

Red-throated Divers breed on isolated Scottish lochs, before heading out to sea to spend the winter. The bird on the right is in breeding plumage, the other in non-breeding garb.

Little Grebes, also known as Dabchicks, are the smallest member of their family, and are easily overlooked. They are best detected by their distinctive, high-pitched call, a repeated series of notes.

Grebes are mainly birds of fresh water, especially during the breeding season. Two species, the **Little** and the **Great Crested**, are common British breeding birds, while the other three, the **Slavonian**, **Black-necked** and the **Red-necked**, are rarer and more localized. All grebes have ornate breeding plumage and impressive courtship displays, during which they 'dance' in the water, rising up almost vertically or pursuing each other in pair-bonding movements, sometimes offering each other pieces of weed as a token of their affection. In winter, all grebes apart from the Little Grebe may also be found on the sea, usually close inshore. All five species dive for food throughout the year. During the breeding season, Great Crested Grebes often cover up their eggs with weed to deter predators, a habit that leads to the pale eggs becoming stained a greenish colour. Once hatched, the young will frequently hitch a ride on their parents' backs – always a memorable sight.

Slavonian Grebes are confined as a breeding bird to Scotland, but, in winter, can be found around many of our coasts.

SEABIRDS (SHEARWATERS, PETRELS, GANNET, CORMORANTS, SKUAS, AUKS)

This motley collection of different families have one thing in common: they are generally found either around our coasts or out at sea (though in recent years the Cormorant in particular has ventured inland to breed).

Shearwaters and **petrels** are 'true' seabirds, spending virtually the whole of their lives at sea and coming ashore only to breed. Four species breed in Britain and Ireland: **Manx Shearwater**, **European Storm-petrel**, **Leach's Storm-petrel** and **Fulmar**. The first three usually return to their island breeding colonies at night, to avoid predators such as gulls, and, as a result, can be hard to see. Manx Shearwaters, as their name suggests, glide on stiff wings low over the sea. They can be observed doing so either from the coast or from a boat, especially during the evenings when they gather offshore before returning to their breeding colony. The two species of storm-petrel are more pelagic in their behaviour, and are usually seen from land during onshore gales in autumn only, when they may appear in quite large numbers. The European Storm-petrel is a tiny bird with a weak, fluttering flight, while its larger cousin Leach's Storm-petrel has a more deliberate flight action, often said to be like that of a nightjar, with flaps punctuated by long glides. The Fulmar superficially resembles a gull,

Manx Shearwaters are virtually confined as a breeding bird to Britain and Ireland: 90 per cent of the world's population breeds here. They forage for food at sea, returning to their breeding colonies by night to avoid predators.

TOP TIP

Pelagic trips
'Seawatching' – observing from a prominent coastal headland – is a good way to see seabirds. To watch them in their natural element you could go on a 'pelagic': a boat trip out into the ocean, where you may see species such as shearwaters and petrels.

but a closer look reveals the characteristic 'tube nose', which marks it out as a true seabird. In flight, its stiff wings and superb aerobatic ability make it a joy to watch.

The **Gannet** is Europe's largest and most impressive seabird, especially when seen at its huge nesting colonies off the Welsh and Scottish coasts. It is superbly adapted for aerial diving, plummeting from a great height into the water in order to catch fish. At its breeding colonies, you can also watch the birds displaying, squabbling and fighting over their tiny territories.

The **Cormorant** and **Shag** were both originally coastal species, and, although the Shag remains so, the Cormorant is now a common sight inland, especially on rivers and gravel-pits. Both species are accomplished divers for fish, but because they do not have waterproof feathers they must stand around afterwards drying their wings in a characteristic pose. Both are also communal breeders – Shags on cliffs with other seabirds, and Cormorants in a variety of places, including trees, where they build untidy nests out of sticks.

Skuas are related to gulls and terns, but are more coastal in distribution than most of their relatives. Both British breeding species, **Great** and **Arctic Skuas** feed by stealing from other birds, a habit known as kleptoparasitism. At seabird colonies they can often be seen chasing birds such as Kittiwakes or terns, in order to get them to regurgitate and drop their food. If you venture close to a skua breeding colony, watch out for another unpleasant habit – that of attacking intruders by flying towards them at great speed. It can be a frightening experience. Outside the breeding season, skuas may also be seen offshore.

The **Auk** is the northern hemisphere's equivalent of the penguin and has many similar characteristics, though it has not lost the ability to fly. Generally, though, auks are rather ungainly fliers, and also unsuited to land. Their real home is underwater, where they can dive to great depths in search of food. There are four British breeding species: **Guillemot**, **Razorbill**, **Puffin** and **Black Guillemot**. Guillemots and Razorbills nest on steep sea-cliffs, laying a single egg on a narrow ledge; Puffins make burrows; while Black Guillemots breed in holes or crevices. Outside the breeding season, all auks become more pelagic in behaviour, heading out to the open sea. Occasionally, a storm will drive them onto the coasts or even far inland, in what is known as a 'wreck'.

The Arctic Skua is a fearsome predator, often stealing food from other birds as they pass by in mid-air.

With their multi-coloured beak and amusing habits, Puffins are surely our best-loved seabirds. In winter, Puffins head away from their breeding colonies and out to the open sea.

Europe's largest species of heron, the Grey Heron is a familiar sight as it stands patiently waiting for its aquatic prey to appear.

HERONS AND EGRETS

The three British breeding species of heron and egret, the Grey Heron, the Bittern and the Little Egret, are representatives of a huge family of long-legged waterbirds found mainly in warmer parts of the world.

The **Grey Heron** is one of our largest flying birds and is a familiar sight either overhead or standing stock still by the side of a lake or river, hunting its aquatic prey. Herons fish mainly by stealth, patiently waiting and watching, then attacking with a swift stab of their sharp, powerful bill. They nest in large colonies called heronries, and are very early breeders, often starting in January and having young in the nest by March.

The **Bittern** is one of our shyest and most elusive birds. Dwelling in reedbeds, it rarely shows itself, only occasionally appearing at the edge of the reeds, or seen briefly in flight as it travels the short distance from one part of its territory to another. In winter, Bitterns can be more visible, especially during harsh weather.

The **Little Egret** is a very recent colonist as a British breeding bird, and, like many of its relatives, it nests colonially in trees. However, you are much more likely to see it hunting for fish and aquatic invertebrates at low tide in coastal harbours or even areas of inland water near coasts. At high tide, Little Egrets come together to roost, sometimes in quite large flocks.

The Bittern's secretive habits and cryptic plumage mean that it is hardly ever seen, apart from occasional short flights.

DUCKS, GEESE AND SWANS

Ducks, geese and swans together make up a large and diverse group of birds generally known as wildfowl or waterfowl. These species have long been associated with humans, either as domesticated creatures or for hunting. Hence many species have developed an understandable wariness of human beings.

Ducks

Twenty-one species of duck are regularly seen in Britain, and can be put in different categories depending on their behaviour and habits.

First are the 'dabbling ducks', so-called because of their habit of feeding by dabbling, that is ducking their heads under the surface of the water, or occasionally up-ending to take food. This category includes some of our more familiar species – the **Mallard**, **Shoveler**, **Wigeon**, **Pintail**, **Gadwall** and **Teal** – and the much scarcer **Garganey**. Each feeds in slightly different ways. For example, the Shoveler sweeps its bill through the water, filtering out tiny morsels, whereas the Wigeon usually graze on land, feeding on grass. Gadwall often accompany Coots, and appear to take advantage of the fact that the Coot stirs up the water when

The Teal is our smallest duck, but its delicately marked plumage and attractive appearance make it many people's favourite. They are shy, wary birds, that usually take flight when they are disturbed.

Mallards are dabbling ducks, feeding by up-ending and taking prey from just beneath the surface of the water.

it dives, and brings morsels of food to the surface that the Gadwall then feeds on. Pintail, Teal and Garganey are generally shy, often flying when they detect human presence. Mallard, by contrast, will generally allow a close approach – indeed most people's first experience of 'bird behaviour' is when, as children, they go to feed the ducks.

Another category is the so-called 'diving ducks': the **Tufted Duck**, **Pochard** and **Scaup**. The first two species are common in Britain, especially during the winter when the breeding population is far outnumbered by immigrants from the north and east. Both usually live on inland waters and dive for food, often going quite deep underwater. The Scaup is a more sea-going species, though it can also be found on reservoirs and large gravel-pits, especially those near the coast.

Tufted Ducks are our commonest diving duck. Males (right) are much more strikingly marked than the browner females. They obtain their prey by diving several metres below the surface of the water.

Another group of diving ducks is sometimes referred to as 'sea ducks', and includes the **Goldeneye**, **Eider**, **Common** and **Velvet Scoters**, and the **Long-tailed Duck**. These ducks, as their group's name suggests, are generally found around the coasts, where they often form large flocks offshore, sometimes consisting of several different species, all diving for food. Goldeneye and Eiders can also be seen courting in early spring, throwing their heads back in display and uttering extraordinary calls

The Eider is Britain's commonest and most widespread sea duck, found off many of our coasts, especially in northern Britain. Eiders feed on crustaceans and molluscs taken from just beneath the water surface.

Three other species are called 'sawbills' because their bills have a serrated edge. These include the **Red-breasted Merganser**, **Goosander** and the **Smew**. Like the sea ducks, they also dive for their food. Red-breasted Merganser and Goosander both breed in Britain, generally on fast-flowing upland rivers or lakes, although in winter, mergansers are usually found near the coast or offshore. The Smew is a winter visitor and prefers gravel-pits, though it is a shy bird and often vulnerable to disturbance.

Finally, there are three 'miscellaneous' species: the **Mandarin Duck**, **Ruddy Duck** and **Shelduck**. Both Mandarin and Ruddy Ducks are introduced species, and each displays some fascinating behaviour, especially during courtship displays.

Shelducks are not really ducks at all, but an intermediate between ducks and geese. They are generally found around the coasts, especially on marshes and estuaries, where they may gather in large flocks.

Male Red-breasted Mergansers perform an extraordinary display to the female, lifting the head and throwing it back over the neck.

The Mandarin Duck was originally introduced to south-east England from its native China and Japan. It nests in holes in trees, from which the young must leap several metres to reach the ground.

White-fronted Geese are so called because of the small white patch on their face – 'front' is an old word for 'forehead'. They are winter visitors to Britain, usually seen in large, noisy flocks.

Our commonest member of its family, the Canada Goose, was originally introduced from North America in the 17th century. In recent years it has increased its population dramatically, causing problems in some areas where it dominates other species.

Geese

Wild geese are some of the best-loved of all British birds, especially when they gather in huge flocks in winter. Yet another species, the introduced Canada Goose, is one of the most loathed of all birds.

Eight species of goose regularly visit Britain, of which four are collectively known as 'grey geese'. This category includes **Greylag**, **White-fronted**, **Pink-footed** and **Bean Geese**. The latter three species are all winter visitors to Britain, arriving in autumn from their breeding grounds to the north and east, and spending the winter months feeding in large, noisy flocks, generally on farmland but also on coastal marshes. The Greylag Goose breeds in northern Britain, but, in recent years, a large feral population has established itself in the south. Geese tend to be creatures of habit, feeding in their favourite areas by day and, as night approaches, gathering to roost nearby, often near water.

Two other species of goose, **Barnacle** and **Brent**, are smaller and more distinctive in appearance. Barnacle Geese have similar habits to the grey geese, but, like the Greylags, a feral, non-migratory population is now at large in southern Britain. Brent Geese prefer coastal marshes and estuaries, where they gather to feed on eel-grass and other vegetation.

The two remaining species are both introduced: **Canada** and **Egyptian Geese**. Both were originally brought here as ornamental wildfowl in the grounds of stately homes, but have since escaped and spread, and are both now firmly established as British breeding species. Canada Geese form large, noisy flocks, and can sometimes drive away other species of wildfowl.

Swans

Three species of swan are found in Britain: one resident, the Mute Swan, and two primarily winter visitors, Bewick's and Whooper Swans (though Whooper does breed occasionally in northern Scotland).

Mute Swans are one of our most familiar birds, and also our largest and heaviest bird. Although they often allow quite close approach, they can be dangerous, particularly to small children, though the oft-quoted remark that they can 'break a man's arm' is a myth. Mute Swans generally pair for life, and return to breed at the same site each year, building a huge nest out of twigs. Sadly these nests are often accessible to human vandals and animal predators, though if the pair is successful they usually manage to raise a large brood of cygnets.

Whooper and **Bewick's Swans** arrive in large flocks each autumn. Whoopers migrate from Iceland at incredibly high altitudes, while Bewick's make an even longer journey here from Siberia. Both enjoy the benefits of Britain's mild winter climate, where plenty of food is available. They usually gather in traditional sites, though small flocks can be found feeding in fields away from these. Dusk is a good time to see them in large numbers, as they gather together to roost for the night.

Mute Swans are the only one of the three British swan species to breed here. Unlike Bewick's and Whooper Swans, they are resident.

The White-tailed (or Sea) Eagle went extinct as a British breeding bird in the early 20th century, but has now been reintroduced to north-west Scotland, where it feeds mainly on fish, sometimes even following trawlers for food.

The Osprey was also driven into extinction, but re-colonized Scotland of its own accord from the early 1950s. It has recently also bred in England. It is a summer visitor, spending the winter in Africa.

RAPTORS

Raptors, or day-flying birds of prey (excluding owls), are a diverse group that includes some our best-known and most magnificent species, such as **eagles**, **hawks** and **falcons**. Fifteen species breed in Britain, though a number of these have a very restricted range and are unlikely to be seen without special effort to do so.

Starting with the largest raptors, two species of eagle breed here: **Golden** and **White-tailed**. Golden Eagles are mainly confined to Scotland, though an occasional pair breeds in the Lake District of England. These are magnificent birds, which cover a vast territory and are never easy to find. In a suitable area, look out for them soaring on huge wings over crags and mountains, though bear in mind that they tend to nest at lower levels than you might expect, to avoid having to carry prey upwards. Like all large birds of prey, they make use of thermals and are more likely to be seen soaring in warmer weather. The White-tailed, or Sea Eagle, went extinct as a British breeding bird in the early 20th century, due to persecution. Since being reintroduced, the species is now thriving on the rocky shores of western Scotland and the outlying islands. Never easy birds

to see, they regularly follow fishing trawlers, consorting with the much smaller gulls to take a share of the fish thrown overboard.

The **Osprey** is the other great fish-eating bird of prey, and has evolved to catch fish in its huge talons, diving down to the surface of a lake in order to do so. As migrants, Ospreys are regularly seen in southern Britain on their journeys to and from Africa, sometimes hanging around wetlands for days or even weeks. They often perch on prominent posts or trees.

Another group of large birds of prey is the **buzzards**. The **Common Buzzard** is our commonest large raptor, and can often be seen soaring high over woods on a fine day, especially in western Britain. Sometimes mistaken for an eagle, its wing-shape is in fact quite distinctive: it usually holds its wings bowed with the tips upwards, and occasionally even hovers rather clumsily. Its much rarer relative, the **Honey Buzzard**, is a shy bird that feeds mainly on the grubs of wasps and bees. A summer visitor, the best chance of seeing them is to visit known breeding sites in late May when the birds are displaying. Unlike the Common Buzzard, they generally soar on flat wings. **Harriers** and **kites** are superficially similar birds, with long wings and a habit of flying low over the ground. The **Red Kite** gave its name to the child's toy and you can see

The Honey Buzzard is a very rare British breeding bird, though migrants from Scandinavia may also be seen in autumn. In flight, they tend to show a flatter wing shape than Common Buzzard.

Red Kites were once confined to mid-Wales, but, thanks to an imaginative reintroduction scheme, they are now a regular sight in many parts of Britain.

why, as it is one of our most graceful and acrobatic birds of prey, often seizing food from the ground without landing. Like buzzards, kites will sometimes hover as they look for their prey. **Marsh Harriers** are birds of reedbeds and other wetlands, and like all harriers they fly low on V-shaped wings while hunting. **Hen Harriers** breed on moorland, though spend the winter in wetland areas. **Montagu's Harrier** is a very rare British bird, generally nesting in open farmland, over which it hunts on narrow wings, giving it a buoyant flight.

The two **hawks** are both birds of wooded and forested areas. The more common **Sparrowhawk** has made a comeback since declining due to agricultural chemicals, and is a familiar sight in towns and suburbs. Often merely glimpsed as it passes by, its fast, low flight is designed to surprise its songbird prey. Sparrowhawks also have short, rounded wings and a long tail to manoeuvre through foliage. This gives them a very distinctive flight action out in the open – a short series of flaps followed by a glide. If you are really lucky, a Sparrowhawk may visit your garden and seize an unsuspecting – and unlucky – songbird before your very eyes, then sit on a nearby post and pluck it ready for eating. **Goshawks**, by contrast, are extremely elusive and hard to see,

Britain's rarest breeding raptor, Montagu's Harrier, was named after the 18th-century ornithologist George Montagu. It is a summer visitor to Britain, nesting mainly on arable farmland.

despite their huge size. They are forest dwellers, and the best way to catch sight of them is in late winter and early spring, when, on a fine day, pairs will display above the forest canopy.

The smallest British raptors are the **falcons.** Four species breed in Britain, of which by far the most common is the **Kestrel**. Kestrels are generally seen hovering – a high-energy hunting strategy that enables them to catch their favourite prey of voles. They may also be seen soaring, gliding, or simply sitting on a post or in a tree, resting and watching out for prey.

Marsh Harriers – a juvenile, a female and a male with its complex pattern – have a distinctive way of flying, flapping and gliding, and hovering.

Goshawks are shy birds, rarely seen except in spring, when they display above their forest habitat.

99

Hobbies hunt on the wing for dragonflies, revealing their wonderful aerobatic skills.

The Peregrine Falcon is another raptor that has made a dramatic comeback following the banning of pesticides such as DDT. They are powerfully-built birds, capable of flying at great speeds when hunting prey.

The **Peregrine**, our largest falcon, hunts in an even more spectacular way, flying high in the sky before plummeting down at great speed onto an unsuspecting pigeon or other bird. Peregrines are among the fastest-moving creatures in existence, reaching speeds of well over 240 km/h (150 mph) when 'stooping'. They can also be seen hunting low over marshes in winter and moors in summer, often putting other birds into a blind panic. In recent years, they have moved into city centres and now nest on a number of large buildings, where they sit and survey their territory. Britain's smallest falcon, the **Merlin**, is another opportunistic hunter, which, like the Peregrine, chases and catches its prey with an amazing turn of speed. Finally, the **Hobby**, a summer visitor to Britain, hunts either by hawking for dragonflies (which it grabs in its talons and transfers to its beak while hardly breaking its flight pattern) or searching for flocks of swallows and martins, which it chases and seizes in flight. Look out for hobbies on fine summer evenings when groups often come together to hunt, especially over wetland areas.

GAME-BIRDS

Ten species of game-bird breed in Britain. Five of them were introduced here, a statistic that reflects these birds' importance as objects of quarry or ornament. Two ornamental species are **Golden** and **Lady Amherst's Pheasants**, both of which are highly localized. Their well-known relative, the **Common Pheasant**, was also introduced, probably by the Romans, but for food rather than pleasure. It has since become our most widespread game-bird, and one of the commonest species of all, largely because millions of birds are released each year for shooting. Pheasants are birds of woodland and woodland edge, and are often very approachable, though when surprised they fly away noisily. Females are often more skulking, especially during the breeding season.

The two species of **partridge**, **Grey** (or English) and **Red-legged** (or French), are often seen together, though the introduced Red-legged is usually bolder and more inclined to sit out in the open. Both species can be very wary and with good reason – partridge shooting is a popular sport. Early mornings and evenings are the best time to look for them, especially on the edges of fields

Golden and Lady Amherst's Pheasants were both introduced to Britain from Asia in the 19th century. Despite their gaudy plumage they are very shy birds and rarely seen.

Grey (or English) Partridges are shy birds, and have undergone a rapid decline in recent years. They prefer traditionally-farmed land with hedges and crops where they can hide.

101

near cover such as long grass and hedgerows. Their tiny relative, the **Quail**, is a rare summer visitor to southern Britain. It is hardly ever seen. If you come across it at all, it is usually by hearing its distinctive call, sometimes transcribed as 'wet-my-lips'.

The four members of the grouse family are a contrasting bunch in appearance, habitat and behaviour. The **Red Grouse** is the species most often hunted, and lives exclusively on open moors in northern and western Britain. Red Grouse are always very wary, often sitting tight before exploding in a flurry of wings and disappearing low over the horizon. Listen for their calls to locate them. **Black Grouse** and **Capercaillie** are both birds of the forest, though Black Grouse also live on more open moors. Both species partake in lekking, in which a number of males compete in a communal display in order to attract females. Sadly, because both species are now rapidly declining, it is not advisable to visit their leks, as this can cause disturbance. Indeed, even walking through their habitat is now discouraged. Your best chance to observe their extraordinary and wonderful behaviour is to visit a site where the birds can be viewed from a hide or the road.

The other grouse species, the **Ptarmigan**, is a bird of the Scottish mountaintops. In winter, it moults into a white plumage for camouflage against the snow, while in spring, summer and autumn it goes through phases of grey and brown to match the boulders among which it nests. Ptarmigan can be very approachable, but beware of disturbing them by going too close.

The male Black Grouse displays to rival males and females in a jousting area known as a 'lek'.

RAILS AND CRAKES

There are just five species of rails and crakes found in Britain, showing very different behaviour. Two species, Coot and Moorhen, are relatively open in their habits; while the remaining three, Water Rail, Corncrake and Spotted Crake, are shy and elusive, with the latter two species often proving almost impossible to see.

The **Coot** is such a common and widespread species in Britain that it would be easy to take it for granted. Yet, when watched closely, their behaviour is fascinating, especially during courtship, when males will fight, sometimes to the death. They do so by leaning back into the water and using their feet and claws as weapons. During the breeding season, Coots are one of the easiest birds to observe without disturbance: their nests are easily found, and, once the eggs have hatched, the chicks and adults appear to be well used to human beings. The young often beg their parents for morsels of food, and the pair can have their work cut out keeping the hungry chicks satisfied. On land or in flight, Coots appear relatively clumsy.

Coots are attentive parents, with the adult (front) looking after the young bird for several weeks after leaving the nest.

Moorhens can be aggressive when defending their breeding territory, fighting with beaks and claws.

The **Moorhen** shows very similar behaviour to the Coot, though instead of diving for food it picks items from off or just beneath the water's surface. Moorhens also frequently feed on land, often on areas of damp grass by water. As they walk, they bob their tail and head in a characteristic manner. Moorhens rarely seem to fly, preferring to paddle rapidly across the water or dive into cover when alarmed.

The **Water Rail** is a much more terrestrial bird than its two aquatic cousins, generally hiding away in dense reedbeds or other fringe vegetation around the edge of ponds or lakes. Its long legs do allow it to wade, but it is really designed to squeeze through narrow gaps in reeds, its body laterally compressed to enable it to do so quickly and easily. Water Rails hunt their prey avidly, spearing or seizing it in that long, sharp bill. They rarely fly, generally flapping for a few yards before reaching cover. They do not appear to fear humans, and can give excellent views, especially if you are patient and prepared to wait.

The Spotted Crake is rarely seen, spending the vast majority of its time hidden deep in dense reedbeds.

The **Spotted Crake** looks much like a miniature Water Rail, and inhabits the same dense, semi-aquatic vegetation. However, it is even more skulking, and often the only clue to its presence is its characteristic whiplash call. Amazingly, for such a skulking bird, it is a long-distance migrant, wintering in tropical Africa.

The **Corncrake** was once known as the Landrail, which gives a clue as to its terrestrial lifestyle. Once widespread throughout Britain, it has suffered from the spread of modern farming and is now confined to the extreme north and west of Scotland and parts of Ireland. There, its call may be a ubiquitous 'sound of summer', but seeing the bird presents a far greater challenge. Corncrakes are ventriloquial, so pinpointing their position in long vegetation is almost impossible; it is best to wait patiently and hope the bird eventually shows itself. Early in the season, when the birds arrive back from Africa, is the best time to try to get a good view.

TOP TIP

Calling rails and crakes *Apart from Moorhen and Coot, rails and crakes are always difficult birds to see: the best way is to wait patiently at a known site, with dawn and dusk being the prime times for activity. Listen for their distinctive calls, which indicate their presence.*

The Corncrake is elusive – your best chance of seeing one is in spring when the male birds are calling.

WADERS

There are 30 or so species of wader that regularly breed, migrate through, or winter in Britain. Dividing them up into groups on behavioural factors is bound to be somewhat artificial, but for ease I have chosen a range of categories, some artificial, others based on classification.

Plovers

Plovers are a diverse group of waders, which nevertheless share some obvious characteristics, including a short, straight bill, usually used for picking items of food off the surface of mud or the ground; shortish legs and a characteristic 'stop-and-start' running action; and long wings, often used for epic, migratory journeys.

The two smallest plovers in Britain are the **Ringed** and **Little Ringed Plovers**, both of which have very similar feeding habits. The Ringed, however, tends to be a bird of coastal areas, often found in the company of other small waders, though preferring to feed singly. The Little Ringed Plover colonized Britain between the First and Second World Wars, using newly dug reservoirs and gravel-pits as nesting areas. Both species have very interesting breeding habits, including a famous 'distraction display', in which the bird will call plaintively while dragging its wing along the ground to draw predators (or human beings) away from its nest or chicks.

The Ringed Plover is primarily a coastal bird, though many pairs do breed inland, by the side of lakes, reservoirs and gravel-pits.

Two larger species, **Grey** and **Golden Plovers**, also form a 'pair'; however, despite their similarity of appearance, they have rather different habits. Grey Plovers are birds of coastal marshes, usually seen singly or in loose groups; whereas, outside the breeding season, Golden Plovers form flocks, often with Lapwings. During the breeding season, Golden Plovers will also perform the distraction display to ward off predators.

The **Lapwing** is many people's favourite wader and with good reason: as well as being stunningly beautiful, it also displays some quite distinctive behaviour. Outside the breeding season, Lapwings form huge flocks, gathering on farmland or coastal marshes to feed. When feeding, they adopt a characteristic plover trait, taking a few rapid steps forward, pecking briefly at a morsel of food, then running forward once again, constantly on the lookout for more to eat. When disturbed, Lapwing flocks rise into the sky, calling in alarm. When courting, Lapwings perform a wonderful courtship display, tumbling through the air like acrobats. Their young are, like all wading birds, precocial (able to leave the nest within a few days of hatching), and cryptically coloured to avoid being caught and eaten.

The final representative of the plover family is the **Dotterel**. Confined as a breeding bird to the highlands of Scotland, it is occasionally seen in southern Britain as a migrant in spring, on its way back from African winter quarters. Flocks (or 'troops') of Dotterel are creatures of habit, often turning up in the very same field, on virtually the same day, from spring to spring. At their breeding grounds, they live up the origin of their name (which means 'fool' or 'dupe') by being very tame, often approachable to within a few metres (though this is not advisable when the birds are breeding).

The Lapwing's erect crest and glossy plumage make it one of Britain's most attractive waders. In winter, they form large flocks and are often seen on coastal marshes and estuaries.

The Dotterel is one of the very few birds in which the female takes the lead in courtship, and sports a brighter plumage than the male.

Britain's smallest wader, the Little Stint is barely larger than a Blue Tit, yet undertakes an epic migratory journey from the Arctic to Africa.

Knots are sociable birds, often gathering in vast flocks to feed at low tide or roost at high tide.

Small waders

This is a diverse group that has little in common behaviour-wise, but the birds are of a similar size.

The smallest of all are two species of **stint**: the **Little** and **Temminck's**. These tiny waders are most often encountered on migration, as they feed frantically to build up fuel before the next leg of their journey between the Arctic and Africa. They may be accompanied by another long-distance migrant, the **Curlew Sandpiper**, with distinctive downward-curving bill. Like their much commoner relatives, **Dunlin**, **Knot** and **Sanderling**, these birds are consummate travellers.

Dunlin and Knot gather in huge flocks to feed and roost, crowding together for safety against predators as the waters rise at high tide. In flight, flocks of Knot appear to be controlled by some unseen hand, as they twist and turn through the air with extraordinary manoeuvrability. Sanderling gather in much smaller groups on the tideline, racing away from the incoming water like little clockwork toys, their legs going like the clappers.

Two distantly related species, **Purple Sandpiper** and **Turnstone**, are also birds of the coast, sharing the habit of roosting and feeding on rocky shores, where they use their short, powerful bills to pick up invertebrate prey. Three other species,

Common, **Wood** and **Green Sandpipers**, are usually associated with fresh water, though, on migration, they also occur near the coast. Common Sandpiper invariably bobs up and down when feeding, a useful identification point (though the Green can do the same thing). Green Sandpipers tend to sit tight until flushed, then fly away noisily, whereas Wood Sandpipers behave more like a small 'shank', wading in deeper water with an elegant movement and action.

Common Sandpipers have a characteristic 'bobbing' movement, jerking their tail up and down when feeding.

Medium waders

There are four medium-sized waders that are often encountered in a range of habitats and locations. The three 'shanks', the **Redshank**, **Greenshank** and the **Spotted Redshank**, are all long-legged waders with fairly long bills, which usually feed on mud or near the edge of water. Redshank are the classic 'all-purpose' wader, adapted to a range of habitats, especially outside the breeding season. Their habit of taking flight and calling in alarm as soon as they are approached has earned them the nickname 'sentinel of the marsh'. In the breeding season, they call incessantly, often perching on fence posts to get a better view of danger. Greenshank and Spotted Redshank are both rather specialized feeders. The Spotted Redshank's long legs and bill enable it to wade quite deep into the water, making it more like a godwit in habits. The odd man out is the

TOP TIP

Tides and waders
A thorough knowledge of tide times is vital if you want close-up views of many species of coastal wader. Get to a roost site a couple of hours before high tide, and watch as the rising waters push the birds closer towards you.

Ruff, which can bear a superficial similarity in structure, build and habits to the Redshank, especially outside the breeding season, when it shares the same habitat. Ruff also feed in drier areas, however, such as ploughed fields.

In the breeding season, the Ruff's behaviour is radically different from that of the shanks. Males adopt a splendid head-dress and gather in leks to try to woo the females. As with all lekking species, Ruffs let the female do all the work of incubating the eggs and rearing the chicks.

Large waders

This motley collection of birds includes the two godwits, Curlew and Whimbrel, as well as oddities such as the Stone Curlew, Oystercatcher and Avocet. The two species of **godwit**, **Bar-tailed** and **Black-tailed**, are both large, long-legged wading birds characteristic of coastal wetlands. They generally gather in flocks, Black-tailed godwits feeding methodically in deep water, and Bar-tailed preferring sandy shores and mudflats.

The **Curlew** and **Whimbrel** both have long, decurved (downward-curved) bills, which they use to probe into mud or soil to find invertebrate food. During the breeding season, both have delightful display flights, uttering their haunting calls as they fly overhead. In winter, Curlews may gather in quite large flocks on estuaries or mudflats, or to feed in flooded fields. Whimbrel migrate to Africa, often stopping off at coastal sites to refuel.

The **Stone Curlew** is not a curlew at all, but a member of an African family known as the 'thick-knees' due to their peculiar anatomy. It is a bird of dry farmland and heath, and its large, staring eyes reveal that it is mainly a crepuscular (twilight) feeder. Stone Curlews are brilliantly camouflaged, especially during the day when they crouch in furrows, and are often only visible when they walk forward on their long legs.

The **Oystercatcher** is a classic bird of coastal areas, found in a variety of sandy and muddy habitats. It can form huge flocks, especially at high-tide roosts, but when breeding, frequents different habitats, including the edge of lochs and rivers, or grassy fields, sometimes far inland.

Oystercatchers are found in a wide variety of coastal and marine habitats – their black-and-white plumage and orange bill make them unmistakable. Despite their name, they feed mainly on mussels.

The **Avocet** is perhaps Britain's most elegant wading bird. It has an extraordinary upcurved bill, which is used in a unique way. Instead of poking, picking or probing the mud, the bird sweeps its bill from side to side to filter out tiny aquatic organisms. Avocets often feed in groups, but, in the breeding season, they can get very territorial and aggressive, especially if an intruder comes near.

Snipe and Woodcock

Snipe, Jack Snipe and Woodcock are three species with long bills, short legs, cryptic plumage and similar feeding habits.

The **Snipe** is generally found in damp grassy areas, wet meadows or marshes, feeding close to cover and probing its huge bill deep into the mud. It is easily alarmed, flying away fast and high on rapidly beating wings. During the breeding season, Snipe perform an amazing 'drumming' display, in which they fly high into the sky before plummeting earthwards, making a strange noise by vibrating their tail feathers.

Its smaller relative, the **Jack Snipe**, is far less easily seen, usually sticking very close to cover. One way to tell the two species apart is the Jack Snipe's characteristic bobbing action, as if its body is mounted on springs. Jack Snipe stay put almost until they are trodden on, before flying away very fast and low to take cover farther on.

The **Woodcock** is, as its name suggests, a bird of woods and forests, where it can be very difficult to see. The birds generally sit tight, relying on their cryptic plumage for protection. Your best chance of observing Woodcock is during the 'roding' display on a fine evening in spring or early summer, when they fly around the tops of trees flicking their wings and calling.

The Common Snipe uses its long, straight bill to probe into soft mud to obtain food.

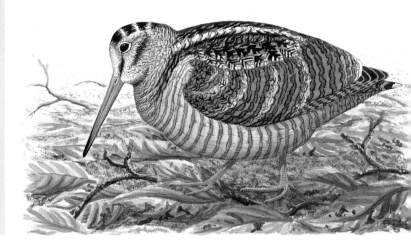

The Woodcock is one of our most secretive waders, spending most of its life hidden in the undergrowth on the forest floor.

GULLS AND TERNS

Seven species of gull and five terns regularly breed in Britain, while several other species occur on passage or as regular winter visitors. There was a time when **gulls** – popularly called seagulls – were largely ignored by birders, perhaps because they are common and familiar, but also because they exhibit a bewildering range of different plumages, which can be confusing. Yet gulls are one of the easiest groups of birds to study, and close attention really does pay off.

The species most likely to be encountered, especially inland, is the **Black-headed Gull**, which has adapted so well to living alongside humans that it is now a familiar sight in a whole range of habitats, including gardens. Other species, such as the **Common**, **Herring** and the **Lesser Black-backed**, are also increasingly found inland, with the two larger species even nesting on city roofs. Watch out for all four species flying into their roosts, especially on winter evenings, when the numbers involved can run into thousands.

Nesting colonies of gulls are also fascinating to watch, with every kind of breeding behaviour, including complex courtship displays, mating, territorial squabbles and, of course, the raising of chicks. It is an avian soap-opera, free for anyone to enjoy.

Larger gulls, including Herring and **Great Black-backed Gulls**, often hang around on the edge of colonies, ready to nip in and grab an egg or a chick that has been left alone for a moment or two. Watch also for parent

Despite its name, Common Gull is not the most common or widespread member of its family in Britain. It is often found inland, especially in winter.

Great Black-backed Gull is the largest member of its family in Britain, and also one of the most marine of the gulls – rarely seen far away from the coast.

The Kittiwake gets its unusual name from its call, which really does sound like someone yelling 'Kittiwake! Kittiwake!'

TOP TIP

Gull and tern colonies
If you want to observe breeding behaviour at close quarters, the best place is a colony of gulls or terns. Because they nest so close to each other, there are many opportunities for interaction between individuals. Colonies are also a great place for stills or video photography.

birds returning to their young. They go through ritualized bonding before the adults finally regurgitate their catch as a meal for the hungry chick.

The **Kittiwake** is one of the most marine of all gulls, nesting mainly on rocky cliffs, though also sometimes choosing warehouses on docksides. They are sociable birds, uttering the cry that gives them their name.

Gulls can also be watched feeding, either at rubbish dumps or when following a fishing boat to catch cast-offs. Looking closely through a feeding flock is a good way to pick up rarer species such as **Glaucous** and **Iceland Gulls**.

Two other species of gull are regular passage migrants or visitors: the **Mediterranean** and **Little Gulls**. The former species, though rarer, also breeds here in small numbers. Little Gulls have a very buoyant, rather tern-like flight.

Terns are, as one writer once said, 'gulls that have died and gone to heaven.' With their buoyant flight and graceful appearance, they really do put clumsy gulls to shame. **Common Terns** are now often encountered inland, either on passage or

breeding, especially on islands in gravel pits and around reservoirs. Like gulls, they have complex courtship rituals. Their close relative, the **Arctic Tern**, has a much more coastal distribution and nests in large colonies, usually on offshore islands. If you make the mistake of getting too close, they will often mob you, even drawing blood with that dagger-like bill.

Sandwich Terns are also birds of the coast; they are more gull-like in appearance than their relatives. **Little Terns** are delightful birds, which nest on shingle beaches around our coasts. Like other terns, they can be seen hunting for food as they dive into the sea.

In recent years, the Mediterranean Gull has spread north and west to become a regular, though still scarce, British breeding bird. It usually breeds in colonies alongside its more common relative, the Black-headed Gull.

This juvenile Common Tern (right) is begging for food from its parent, even though it is big enough to find food for itself.

115

Wood Pigeons are often overlooked because they are so common – yet close observation reveals interesting behaviour, especially during the breeding season.

Feral Pigeons come in all shades, patterns and colours, yet are all unmistakably descendants of the wild Rock Dove. In spring, look for their very public courtship displays.

PIGEONS AND DOVES

There really is no clear difference between pigeons and doves, although doves tend to be smaller and more delicate. There are five British species (six if you count Feral Pigeon), three of which are almost ubiquitous: the Feral Pigeon, Wood Pigeon and Collared Dove. The Feral Pigeon's ancestor, the Rock Dove, is a fairly localized bird confined to northern and western Britain, while the Stock Dove is widespread but rarely seen, and the Turtle Dove, the only migratory member of its family in Britain, is much scarcer than in the past.

Paradoxically, the **Feral Pigeon** is a common British bird. However, it is virtually ignored by birders thanks to its dubious origin as a domesticated species. If you want to observe bird behaviour at close quarters it is one of the best species to choose: widespread, used to humans and with an interesting range of behaviour to watch. In early spring, look out for males performing their courtship display to nonchalant females. To see the wild ancestor of this bird, the **Rock Dove**, you will need to travel to outlying islands and headlands in north-west Scotland, where the last remaining pure-bred birds still hold their own. These behave very differently from their feral descendants, flying away as soon as you get near.

The **Wood Pigeon**, originally a bird of woodlands and farmland, has adapted very well to living alongside humans in cities, towns, villages and gardens. It is not the most graceful of birds, but like all successful species, has learned to be versatile in both its feeding and breeding habits.

The **Collared Dove is** a fairly recent colonist from Europe, and has become a familiar bird of towns and suburbs, especially well-wooded ones. Collared Doves often visit bird tables, though like all pigeons, they remain constantly wary, always on the lookout for danger.

The **Stock Dove** may prove to be a hard bird to find. They keep themselves to themselves for much of the year, though on fine days in spring look out for pairs displaying over woodland.

Outside the breeding season, Wood Pigeons, Feral Pigeons and Stock Doves may join together in flocks to feed, especially on farmland.

The smallest British pigeon or dove, the **Turtle Dove**, is also a very shy bird, best detected by its purring call in May and June, after it has arrived back from Africa. Turtle Doves often prove elusive, and are more likely to be seen easily on autumn migration, when they leave woods and heaths and feed on farmland and in coastal areas.

Our smallest species of dove, the Turtle Dove is also the only long-distance migrant in its family, heading back and forth from Africa each spring and autumn.

117

OWLS, CUCKOO AND NIGHTJAR

This miscellaneous group of non-passerine birds includes our largest family of nocturnal birds, the owls, with five British breeding species; another nocturnal bird, the Nightjar; and a summer visitor that lays its eggs in other birds' nests, the Cuckoo.

Owls are difficult to observe, largely because they are either nocturnal or crepuscular in habits, and even those species that do fly by day may be hard to find. Once seen, they are unmistakable: a combination of shape, forward-facing eyes and behaviour marks them out as unique. Owls have adapted to fit

specific habitats, and, as a result, two species are rarely seen together. Our commonest species, the **Tawny Owl**, is also one of our most secretive. It is highly sedentary, spending most of its life in the same small territory, which it gets to know very well indeed. This is essential for a bird that is almost entirely nocturnal. As a result, the best way to find this species is to listen for its characteristic call in late winter or early spring, when birds are marking out their territory prior to breeding. Once you have found a territory, look for suitable nesting holes, usually halfway up a mature tree, and hope that you are lucky. Another way to find Tawny Owls, especially outside the breeding season, is to search for a roosting bird. During the daytime, Tawny Owls sit tight, usually in a hollow in a tree-trunk.

Although one of the most nocturnal members of its family, the Tawny Owl can be seen at a daytime roost in autumn or winter.

Barn Owls exploit a quite different habitat: open farmland, ideally of a traditional nature, with old buildings that they can enter and make their nest. They are often seen at dawn or dusk, hunting like a white ghost over fields and marshes in search of their favourite prey – voles. With their soft plumage designed to allow silent flight, Barn Owls are especially vulnerable to getting wet, so if it has rained for a day or two and then stopped that is a good time to look out for them.

If you do find a bird you will marvel at its ability to glide silently across the ground, then plummet to land to catch its prey.

Little Owls are the most diurnal (daytime) owl species (along with Short-eared). They were introduced to southern Britain back in the 19th century, and have become an established and welcome member of our avifauna. Little Owls often perch on the sides of trees (especially oak), or on stumps, fence posts or the roofs of farm buildings, where they sit and wait before plunging to catch their prey of insects, worms and small rodents. They are often found in parks, though an early morning visit is essential if you are to see them before they are disturbed by people and retreat to their hiding-places.

Barn Owls are consummate hunters, catching voles by approaching them on silent wings, before diving down for the kill.

119

Long-eared Owls are essentially nocturnal birds, and are best detected by the calls of their young, which sound like a squeaky gate.

Long-eared and **Short-eared Owls** form a 'species pair', yet have quite different habits. Like the Little Owl, the Short-eared is primarily diurnal, and will hunt low over moors and marshes on long, lazy wings, often looking more like a harrier than an owl. As a result, it is probably the easiest owl species to see well. The Long-eared owl is a complete contrast: almost entirely nocturnal, and very hard to see. However, you may be lucky enough to discover (or find out about) a daytime roost, where up to a dozen birds will spend the daylight hours huddled together asleep.

The **Nightjar** is another difficult bird to see. You have to locate a specific site (usually a lightly wooded heath or young conifer plantation,

The Nightjar is best detected by listening for its distinctive churring call on a fine spring evening, at its heathland breeding sites. At dusk, it emerges to hunt for flying insects, which it catches in mid-air.

mainly in the south of Britain) and wait patiently at dusk – preferably on a fine evening between mid-May and July. If you are lucky, you will witness the incredible sight of this extraordinary bird hawking for insects, flashing the white patches on its wings while uttering its churring call. During the day, Nightjars roost on heather or on the ground and should not be disturbed.

A whole book could be written about the extraordinary behaviour of the **Cuckoo**. Indeed several have been. This species is our only parasitic breeding bird, laying up to 20 eggs in different nests of its host species – a strategy that maximizes its chances of breeding success. Cuckoos parasitize a particular host species (the one in whose nest they were born), the four most common in Britain being Meadow Pipit, Dunnock, Reed Warbler and Robin. To do so, the female ejects one of the host's eggs before depositing her own. Once hatched, the Cuckoo chick grows rapidly, ejecting any remaining eggs or chicks. The unsuspecting host parents feed it frantically until it becomes far too big for its tiny nest. Meanwhile, the Cuckoo's real parents depart for Africa in June or July without ever seeing their offspring, which, having fledged, manage to find their way south by themselves. Watch out for Cuckoos in late April or early May, when males have just arrived back and are easier to see as they sing from prominent places to attract a female.

The Cuckoo's onomatopoeic call is surely the classic sound of the British summer. Its reputation for laying eggs in other birds' nests has made it one of our best-known birds.

121

PARAKEET, KINGFISHER AND DIPPER

This is another motley collection of birds – two of them are linked by their gaudy colours, and the other is a passerine that thinks it is a waterbird.

A recent addition to our avifauna, the Rose-ringed Parakeet is now a familiar sight in parts of suburban London and the south-east. A noisy, sociable bird, it gathers in huge flocks to roost at night.

Parakeets should have no place in a book about British birds, but they have somehow become an established part of our avifauna during the past three decades or so, since a flock (or more) of birds escaped in the London suburbs. These flocks have grown in size to a population of more than 5,000 birds, which have easily adapted to their new surroundings.

Originally from northern India, the **Rose-ringed** (or Ring-necked) **Parakeet** is a highly adaptable species. It is able to withstand extreme cold and to exploit artificial and natural food resources, including food put out by humans. The birds' favourite habitat is a large wooded park, where they nest in holes in trees (possibly threatening native hole-nesting species, such as the Jackdaw, Stock Dove and Starling). They are easy to see, thanks to their habit of flying in flocks while uttering noisy, high-pitched, contact calls. If you can get up close, you can observe the full range of feeding and courtship behaviours that makes the parrot family such an interesting one. At dawn and dusk, the parakeets fly overhead in flocks of up to 50 or more birds, en route to a communal roost.

Kingfishers are shy birds, but may sometimes be watched at a regular site, where they may give unforgettable views, as they dive beneath the water to catch small fish.

The **Kingfisher** is even more colourful than the parakeet; indeed, it has no rival for the position of our most colourful native species. It is a much smaller bird than many people expect, and it can also be elusive, often only seen as it flies away in a flash of blue and orange. However, find a regular site and you may be lucky enough to get excellent views of the bird feeding by plunging into the water for small fish and other aquatic life. Kingfishers nest in holes in sandy banks. Although the nest itself is hidden underground, the birds can be observed going to and fro.

The **Dipper** is unique. It is a songbird that hunts for its food underwater. Superficially resembling a huge wren, its black-and-white coloration has given it the folk-name of water ouzel. Dippers favour fast-flowing streams and rivers, and, once you have found them, they will provide hours of entertainment as they fly to and fro, perch on rocks bobbing up and down, or plunge beneath the water in order to catch their aquatic food. They nest underneath the banks of the river in a crevice or hole. The young leave the nest before they are fully fledged, and are fed by the parents.

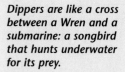

Dippers are like a cross between a Wren and a submarine: a songbird that hunts underwater for its prey.

Green Woodpeckers are quite shy, but will feed on lawns and open grassy areas, searching for insects. This bird is using its specially evolved tail and claws to climb up the trunk of a tree.

TOP TIP

Woodpeckers
Like many woodland birds, woodpeckers are often secretive, and may fly only short distances in a distinctive undulating manner. The best way to locate them is by either their calls, or by listening for their distinctive drumming, most noticeable in early spring.

Dead and dying trees are a good source of food for Great Spotted Woodpeckers. They extract insects and larvae with their strong bills.

WOODPECKERS

Britain has only three native species of woodpecker, compared with ten in continental Europe – but there are none in Ireland. The reason is simply that as poor flyers and largely sedentary species, woodpeckers only spread north and westwards slowly after the end of the last Ice Age. Three pioneering species (plus the now extinct Wryneck) managed to cross the land bridge to England before the sea cut us off from the continent, while none quite managed to reach Ireland before it, too, became an island.

Two out of three British woodpeckers are common and easy to see. Our largest species, the **Green Woodpecker**, is a bird of relatively open grassy areas with scattered trees, such as large

parks, where it can be seen on the ground, feeding on its favourite food of ants. It also visits large, open lawns, but is quite shy and will always be on the lookout, so don't approach too closely. This species drums less often than its 'spotted' relatives, and is best located by its far-carrying laughing call, which earned it the folk-name of 'Yaffle', and is supposed to forecast rain.

Of the two smaller black-and-white woodpeckers, the **Great Spotted** is by far the commonest and most frequently seen. It drums more often than any other woodpecker, and can also be detected by its penetrating 'chip' call. The Great Spotted prefers more dense woodland than the Green, but is also a frequent visitor to gardens, where it will readily feed on bird feeders. It also preys on birds, raiding nests and nestboxes for chicks. In flight, look out for its typical undulating motion.

The smallest European woodpecker, the **Lesser Spotted,** is much more elusive. Indeed, it tends to behave much more like a passerine than a wood-pecker, creeping around the topmost branches of a tree like a Treecreeper or Nuthatch (which it resembles in size, if not appearance). The birds can be easier to see in the early spring when they are calling and drumming; or during winter, when they often tag along on the edge of a tit flock as it passes through woodland in search of food sources. Nevertheless, you very rarely get good views of this species.

Great Spotted Woodpeckers are far more common and more easily seen than the smaller Lesser Spotted. Watch out for them as they hunt for insect food around the trunks and branches of trees.

To see a Lesser Spotted Woodpecker, you need patience. Watch out for them excavating their nests on the underside of dead branches.

SWIFT, SWALLOWS AND MARTINS

Despite a superficial similarity, swifts are not closely related to hirundines, which is the group that includes swallows and martins. Nevertheless, I have treated them together for the sake of convenience, as they share similar behavioural traits.

Swifts are among the most incredible birds in the world. They are the ultimate flying machine, able to stay aloft for months (even a year or more), and landing only to breed. Even then, a Swift will never intentionally land on the ground, as its tiny legs, which are at the rear of the body, mean that it can never get aloft again under its own steam. As a result, Swifts nest in high buildings, where they can perch easily. They return en masse in late April or early May, appearing in

Swifts are the ultimate flying machine: spending virtually their whole lives airborne, where they feed on flying insects. They once earned the folk name of 'devil bird' because of their piercing screams, which were supposed to remind us of lost souls in hell.

huge numbers over reservoirs and marshes where they can find plenty of insect food to refuel after their long journey from tropical Africa. Then they disperse to towns and cities throughout the country, where they spend the evenings chasing each other across the skyline, uttering the screams that gave them the folk-name 'the devil bird'.

Swifts begin nesting in May, but, if bad weather arrives, they will often disappear for days or even a week or more, their young staying torpid in the nest until their parents return and begin feeding them again. The sight of Swifts flying high for insects on a warm summer's evening is one of the most characteristic scenes of summer life, and, when they suddenly disappear in August, the urban landscape seems a poorer place without them.

Of the three true hirundines, the **House Martin** is the Swift's

House Martins build a cup-shaped nest under the eaves of houses, out of tiny balls of mud.

urban and suburban companion. As its name suggests, it has adapted brilliantly to live alongside humans, having originally nested in caves and on the sides of cliffs. The birds arrive back in late April, check out their nesting sites, then often disappear for a week or so to feed on nearby lakes or reservoirs. Once they return, they get down to repairing and rebuilding their old nest, or starting a new one from scratch, using mud collected from a nearby stream, building site or farmyard. They will also readily take to artificial nestboxes – especially useful when a supply of mud is not easily available. They are a wonderful bird to observe as they go to and fro, feeding their hungry young, though their noisy calls can wake you up early. On fine summer evenings (and indeed well into the autumn), they can be observed hawking for insects, sometimes high in the sky.

TOP TIP

Hirundines and the weather
Fine summer evenings are an excellent time to watch swifts and hirundines, as they feed on insects, which they take on the wing. They can also be used to predict the weather: if they fly high, it will be fine the next day; if they fly low, bad weather is on the way.

Sand Martins are gregarious birds, nesting in colonies in river banks. They are the smallest of the hirundines.

Sand Martins are a bird of gravel pits, quarries and riverbanks – in fact anywhere where they can burrow into a sandbank and raise their family. They are a very early arrival, often by mid-March, and like other hirundines they feed over water where there is a plentiful supply of small, flying insects. If you find a colony, it will give you hours of absorbing observation.

The **Swallow** is not only a sign of summer, but also one of our best-loved birds. Its graceful flight and habit of nesting close to human habitation (especially in farmyards) make it a familiar and welcome summer visitor. Swallows will build their nests in buildings and barns, and can easily be detected by their noisy calls. They tend to hunt lower than martins, often feeding alongside farm animals which attract insects. In late summer and early autumn, look out for gatherings of swallows and martins as they perch on telegraph wires prior to migration, filling the air with their contact calls.

Before migrating south to Africa, Swallows and House Martins gather on telegraph wires, uttering their twittering calls.

LARKS, PIPITS AND WAGTAILS

Of these three groups of birds, two are not related (larks and pipits), but resemble each other superficially; and two are related (pipits and wagtails), but may at first sight appear quite different. However, a closer look at these will reveal similarities in structure and behaviour.

There are three species of **lark** found in Britain, two breeders and one wintering. Of these, by far the commonest is the **Skylark**, one of our best-known and best-loved birds, which sadly, in recent years, has undergone a decline in numbers.

Nevertheless, it is still widely distributed in the countryside, in various habitats ranging from lowland farms to upland moors. It is most famous for its extraordinary song-flight, during which the bird rises so high in the sky that it may be almost invisible, pouring its heart out in a continuous song before plummeting down to earth. Once on the ground, it generally lands some distance from its nest and runs the last few metres, making it hard to discover the location of the nest. Outside the breeding season, Skylarks form large flocks, which range over stubble fields to feed. They may also undertake local movements, especially during hard weather in winter.

The **Woodlark**, despite its name, is actually a bird of open heath with a few scattered trees. Once threatened with

The Skylark's song has been justly celebrated by many poets, including Percy Bysshe Shelley, who described the bird as a 'blithe spirit'.

The Woodlark's song is less well-known, though just as beautiful as that of its commoner cousin.

Often dismissed as just a 'little brown job', the Meadow Pipit is a delightful bird, found over most of Britain. In spring, look out for their 'parachute' song-flight.

The Pied Wagtail is one of the classic birds of towns and cities, often feeding on lawns and garden paths.

extinction as a British breeding bird, it has made a remarkable recovery in recent years, thanks partly to the felling of conifer populations and the creation of more open heath. It has a beautiful song, often uttered early in the spring from a song-post on a bush or tree.

The **Shore Lark** is a winter visitor to Britain. As its name suggests, it is found on sandy shores and coastal saltmarshes. It usually forms loose flocks, often associating with Skylarks, and Snow and Lapland Buntings.

Our three breeding species of **pipit** exploit a wide range of habitats. The **Meadow Pipit** is highly adaptable and catholic in its choice of habitat, and is a typical bird of moors, heaths and grassy fields. In winter, it also comes down to the coast, where it may be confused with its much more specialized relative, the **Rock Pipit**. This species is unique among British birds in that it is the only songbird to have an exclusively coastal breeding distribution. The **Tree Pipit** is more a bird of heathland and the edge of woodland plantations, performing its typical song-flight from a high perch, and launching itself into the air before parachuting down again. Finally, the **Water Pipit** is also unique. It is the only British songbird to arrive from the south as a non-breeding visitor. Breeding in the high mountains of Europe, it spends the winter in a range of habitats, including riverside, watercress beds and marshes. All pipits are gregarious birds, often forming loose feeding flocks.

Of our three wagtail species, the **Pied Wagtail** is by far the most widespread and adaptable. It seems to love concrete and tarmac, often being the only bird to walk around this unpromising 'habitat', apparently picking up tiny insects while wagging its tail. Pied Wagtails will also regularly visit garden lawns. They roost in some unusual places, such as trees in the centre of city squares and shopping centres, or factories, where they take advantage of extra warmth provided by industry or retail outlets. In flight, they give a characteristic call.

The Grey Wagtail loves water, and is often found by fast-flowing upland streams, where it may be accompanied by that other upland water specialist, the Dipper. Grey Wagtails nest in crevices, often in stone bridges across rivers and streams.

Yellow and **Grey Wagtails** can be confused on first view, as both have plenty of yellow in their plumage; however, they are structurally quite different. In addition, Yellow Wagtails are a summer visitor, generally seen on marshes and flooded fields (an increasingly scarce habitat), while Grey Wagtails are resident, and associated with running water such as streams or rivers. However, they can also be found at the edges of reservoirs and ponds. Both feed in the typical wagtail manner.

This Grey Wagtail is reacting to its reflection in a mirror, which it thinks is a rival male intent on invading its territory.

THRUSHES AND CHATS

This group of birds includes a dozen familiar and not-so-familiar species, including six 'true' thrushes, and six smaller species that are closely related to each other, and which are also members of the thrush family.

Of the 'true' **thrushes**, the most familiar must surely be the **Blackbird**. Found throughout Britain, apart from some upland areas (where it is replaced by the Ring Ouzel), it is a common and familiar resident in towns, suburbs and the countryside, and is particularly partial to nesting in gardens. Early in the year, listen out for the deep, fluty song of the male, or the angry chattering call as it chases away potential rivals. Blackbirds are highly territorial, and the male will continue singing even when he is feeding chicks. The female is a much less obvious bird, which generally feeds by creeping around in the undergrowth.

The Ring Ouzel is sometimes called the 'mountain Blackbird', as it ecologically replaces its commoner relative in Britain's upland areas. In the absence of trees, males often sing from high boulders.

The **Ring Ouzel** is the upland equivalent of the Blackbird, so is mainly seen in northern and western Britain. It is a summer visitor, and may sometimes be found on migration in lowland habitats, which it tends to visit year after year. Like the Blackbird, it is an excellent songster, perching high on bushes and rocks in order to deliver its song.

The **Mistle Thrush** and **Song Thrush** are often confused, though their size and plumage details are farily distinctive. Behaviourally, too, they differ. The Song Thrush prefers more wooded habitats and gardens, generally singing from the top of roofs; while the Mistle Thrush is a bird of open parkland with scattered trees, which it uses as song-posts. The Mistle Thrush also has the reputation of singing before and during bad weather, which earned it the country name of 'Stormcock'. Outside the breeding season, Mistle Thrushes often gather in flocks, calling as they fly overhead in search of open areas of grass on which to feed. In winter, they defend berry bushes against all-comers. Song Thrushes tend to be shyer and more solitary, leaving gardens for nearby wooded areas.

In autumn, these native species are joined by our two 'winter thrushes', the **Redwing** and **Fieldfare**. In some ways, these are the northern equivalents of the Song Thrush and Mistle Thrush, respectively. Both species travel in loose flocks, sometimes with each other, and raid berry bushes or feed in the open in fields. They also migrate in flocks, calling to each other as they pass overhead, often at night.

Each of our two species of **chat** exploits a slightly different habitat. **Whinchats** are birds of upland areas, such as moors, although on migration they can be seen almost anywhere, while

Whinchats are summer visitors to Britain, and may be seen on passage in spring and autumn in lowland, as well as upland, habitats. They often perch high on a gorse bush or fence, in full view.

Stonechats are more associated with gorse and heathland. However, those birds that do not migrate will spend the winter in more general habitats, such as near reedbeds. Whinchats all migrate to sub-Saharan Africa for the winter.

A close relative, the **Wheatear**, is also a migrant, arriving back as early as March. This is also a bird of moorland areas, though is often found on beaches and other coastal areas during migration, feeding on the ground. Its name has nothing to do with wheat – it is a corruption of an Anglo-Saxon word meaning 'white-arse', which, when you see the bird flicking its tail and wings to reveal its white rump, seems very appropriate.

The Black Redstart colonized Britain as a breeding species around the time of the Second World War, using bomb-sites and industrial areas as breeding sites.

The names of two other small thrushes, the redstarts, also derive from Anglo-Saxon, 'start' meaning 'tail'. The reddish-brown tail is an obvious identification feature. The **Redstart** is a summer visitor, and mainly found in mature, broad-leaved

The Common Redstart is a bird at home in oak woodlands, where you may catch a glimpse of the red tail that gives the bird its name.

woodland, where its song is often the first clue to its presence. It can be a shy bird, but may be seen visiting its nest-hole. Like other chats and thrushes, it may be found in a wider range of habitats on migration. Its rarer cousin, the **Black Redstart**, has more peculiar tastes in terms of habitat. On the continent, it is a bird of rocky slopes and cliffs, but in Britain it prefers to breed in industrial areas such as building sites and even nuclear power stations, having colonized via bomb-sites after the Second World War. In autumn and winter, the birds often disperse to coastal areas, and are generally found near water where they can find insect food.

The final member of the 'small thrushes' is the bird celebrated more than any other by poets and writers – the **Nightingale**. It is a paradoxical bird, one with a stunning song yet a drab plumage and shy, retiring habits. Your best chance of seeing one is when the males arrive back in late April and early May and begin singing to defend a territory and attract a mate, often sitting right out in the open. Once the females arrive and they pair up, the males become incredibly shy, singing from the centre of dense foliage in their woodland habitat or on the edge of heaths. At this stage, they may simply be impossible to see, but they sound wonderful. As you might expect, they do sing mainly at night, though they often give a burst of song during the daylight hours as well.

> **TOP TIP**
>
> **Singing thrushes**
> *Thrushes are amongst the finest singers of all our birds, and often choose a prominent post to deliver their song. Learning the difference between the songs of the three common species is a good grounding in getting to know and appreciate birdsong.*

The Nightingale is justly celebrated for its amazing song – once heard, never forgotten. Although they sing mainly by night, Nightingales also sing by day, especially in spring when they have just arrived back from Africa.

135

ROBIN, DUNNOCK AND WREN

These three familiar garden species are among our more common and best-known birds, and each provides a perfect opportunity to study bird behaviour at close quarters, often from the comfort of your own home.

The **Robin** regularly wins polls of Britain's best-loved bird, yet has some of the nastiest habits of any songbird, with rival males fighting viciously, sometimes even to the death. Once they have established their territory, they will defend it violently, and keep a close eye on the female to make sure she does not stray. Robins nest in a wide variety of places, including bizarre locations such as toilet cisterns, under the bonnet of vehicles, and at the top of drainpipes. They will also readily take to nestboxes. Once hatched, the young may appear in your garden, looking quite unlike their smart, red-breasted parents. In autumn, our native Robins are joined by continental immigrants, which often appear in quite large numbers on the

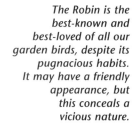

The Robin is the best-known and best-loved of all our garden birds, despite its pugnacious habits. It may have a friendly appearance, but this conceals a vicious nature.

east coast during 'falls' brought about by rough weather. Robins are one of the very few birds that sing all year round (even at night, which often leads to out-of-season claims of Nightingales).

The **Dunnock** has an equally torrid sex-life, in which males follow the female about to prevent another male mating with her on the sly. He will even use his beak to remove a rival's sperm from her cloaca. He is often polygamous, and will spend his time singing to defend his territory – one of the few times the Dunnock will sit out in the open. Otherwise, these birds tend to creep about flower borders like little mice, occasionally venturing out onto the lawn to pick up seeds dropped from a feeder or bird-table.

Wrens are also very territorial, and, like the Dunnock, the male becomes highly visible in early spring, often singing right out in the open, with an incredibly loud song for such a small bird. Male wrens often have to build several nests before the fussy female is satisfied, and chooses one in which to lay her eggs. Outside the breeding season, Wrens hop about unobtrusively in search of insects, and, in harsh winter weather, will often roost in empty nestboxes, with several birds huddling together for warmth.

Dunnocks are often overlooked – watch out for them creeping about at the back of a shrubbery or hedgerow, as they search for tiny insects to eat.

Despite being Britain's commonest and most widespread bird, Wrens are also often overlooked – their small size means they are more likely to be heard than seen. Listen out for their loud, trilling song throughout spring and summer.

WARBLERS

There are 15 regular breeding species of warbler in Britain (including the two 'crests'); they fall into three categories: wetland, scrub, and leaf warblers and crests.

Wetland warblers

The Reed Warbler is a long-distance migrant, making the journey back from sub-Saharan Africa in just a few long-haul leaps.

This category includes five species that are habitually associated with wetland areas during the breeding season, though the actual habitat varies from species to species. The classic 'reedbed' species is the **Reed Warbler**, which, as its name suggests, lives almost exclusively in reeds, where it makes its nest by weaving grasses around the reed stems. It can be heard delivering its distinctive, repetitive song from mid-April, though birds may be reluctant to show themselves, especially in windy weather when they tend to stay low down in the reeds.

Its close relative, the **Sedge Warbler**, shares its reedbed habitat, although it tends to sit on a more prominent perch, such as a small bush, in order to deliver its song. Males also launch themselves into the air in a song-flight, parachuting down to their perch as they sing.

The other member of this trio, the **Marsh Warbler**, is much rarer in Britain, and prefers damp, wooded habitat on the edge of ponds and streams. It is a highly accomplished mimic, and its song may include snatches from the songs of many other birds – not just British species, but those encountered at its African winter-quarters as well.

The **Grasshopper Warbler** is so named after its extraordinary 'reeling' song, which sounds like a cross between an insect and a fishing reel. It, too, prefers a less 'wet' habitat, and is often found in quite dry, bushy areas near the edge of marshy ones. It tends to sing most at dawn, dusk and even through the night, and although not usually easily visible, once discovered, may allow quite a close approach. The final 'wetland' species, **Cetti's Warbler**, is a fairly recent colonist to Britain, and, unlike most other warblers, it is a resident species. Its presence is usually noted

TOP TIP

Identifying warblers
As fast-moving, mainly sombre-coloured birds, warblers are best identified by a combination of their jizz, habitat, and sound. Each spring, it's a good idea to re-acquaint yourself with their different songs and calls, which will make it easier to identify any unusual visitors.

when it sings its incredible, explosive song, once described as being 'like a Wren on steroids'. Occasionally the bird may show itself, and may even give good views. However, it can also be frustratingly hard to locate.

Scrub warblers

This category includes large warblers of the genus *Sylvia*, a largely Mediterranean group of birds, of which five species breed in Britain. The most familiar of these is probably the **Blackcap**. Not only is it a common summer visitor, found in a variety of rural and suburban habitats, including large gardens, but, in recent years, a population from central Europe has also begun to spend the winter in Britain. These birds often visit gardens in search of food. The Blackcap is a fine songster, often compared with the Nightingale, though without the same range and beauty. Garden visitors have become quite adaptable, feeding on bird tables and feeders, though, in summer, the species is mainly a bird of woodland areas. The Blackcap's sibling species, the

The Blackcap was once known as the 'March Nightingale', as it is one of the first spring migrants to return. Its fluty, melodic song is reminiscent of the Nightingale's, though less loud and varied.

Often overlooked, the Garden Warbler is an anonymous-looking and rather shy bird. Its song is faster and less varied than the Blackcap, though very similar to many ears.

The Lesser Whitethroat is a shy and skulking species, often detected by its distinctive, rattling call.

Garden Warbler, is, despite its name, not a very frequent garden visitor. It prefers fairly open woods, but because of its retiring habits, unmarked plumage and very similar song to the Blackcap, it is often overlooked. The song tends to be faster, less varied and lacking the fluty tones of its commoner relative.

The two whitethroats also form a species pair, separated by their different choice of habitat. The **Common Whitethroat** is found in a range of habitats, including heath, hedgerows, farmland and parks, and draws attention to its presence by singing its rapid, scratchy song either from a prominent perch or by launching itself into the air in its song-flight. After the breeding season it may be found feeding on berries in preparation for its long journey to sub-Saharan Africa.

The **Lesser Whitethroat** is far more elusive than its cousin, and its presence may only be detected by its sharp, dry call or fairly distinctive song, emanating from dense scrub or bushes.

The final member of this group, the **Dartford Warbler**, is along with Cetti's, the only truly resident member of its family. It prefers gorse and heathland as a breeding species, often singing prominently from the top of bushes, especially during fine spring

weather. Outside the breeding season some birds disperse to less specialized habitat, including bracken and parkland.

Leaf warblers and crests

The three 'leaf' warblers are well known to be so similar that they were not told apart until the 18th century, when Gilbert White distinguished between the species. In fact, with modern identification techniques and optics the three species are quite straightforward to identify. They also display quite different behaviour.

The **Willow Warbler** and **Chiffchaff** are the two most similar species of the three. As well as their very distinctive songs, they also exploit rather different habitats, with the Willow

The Willow Warbler is our commonest summer visitor, and sings its distinctive, silvery song in most woodland habitats in Britain, arriving back from Africa in April.

The Chiffchaff is another bird named after the sound it makes, and can be heard from as early as March. At this time of year it can be easily seen as it hops around the bare trees, delivering its song.

The Wood Warbler is the largest of the three 'leaf-warblers', and has a beautiful and distinctive song, heard in oak woodlands from April to June.

Warbler preferring heathland as well as mixed woodland, and Chiffchaff often found on the edge of woodland. Both species are active feeders, and on migration can be found in unusual habitats such as coastal areas. Chiffchaffs also winter in considerable numbers, often near water where they can be seen hunting for insects.

The **Wood Warbler** is a much larger bird and has a very distinctive singing behaviour – shivering its wings in time with its delightful song, then flying a short distance to another part of its territory and starting to sing again.

The two 'crests' are very active, tiny birds (the smallest in Europe), constantly on the move in search of insects. Both are found in a variety of woodlands, though **Goldcrests** have a preference for conifers and will often hunt deep inside the foliage and be hard to see. Listen out for their distinctive calls and song, which are often the best way to find them. **Firecrests** tend to move on more quickly from tree to tree than Goldcrests, and are often found in autumn and winter near the coasts, or near water, where a milder climate encourages more insects. Both species will follow flocks of tits in winter.

Our smallest native bird, the Goldcrest is also one of our most delightful, often coming very close to the observer.

FLYCATCHERS

Two species of flycatcher, both summer visitors, breed in Britain. The **Spotted Flycatcher** is the most widespread, found in a range of lowland habitats including rural gardens and woods. Like all members of its family it lives up to its name, sallying forth from a branch or twig to catch small flying insects in its beak, before returning to its perch. Spotted Flycatchers are one of the latest migrants to return, and once here they nest in crevices in walls, open ledges or in tree-forks.

In recent years, like several sub-Saharan migrants, Spotted Flycatchers have drastically declined in numbers, for reasons as yet unknown, but probably due to droughts in Africa.

The **Pied Flycatcher** is, by contrast, a hole-nester, and prefers mature, mixed woodlands, mainly in the western half of Britain. Like its relative, it too, flycatches for food. Both species may turn up in unusual places during migration.

Pied Flycatchers take readily to nestboxes, and may be observed on several RSPB woodland reserves in the south and west of Britain.

The Blue Tit is the commonest and most familiar member of its family, found in virtually every British garden. Its small size, attractive plumage and cheeky habits make it one of the most popular British birds.

TITS, NUTHATCH AND TREECREEPER

This set of woodland species often spends time in close proximity, especially during autumn and winter when they will form mixed flocks comprising several different species, to hunt for insects. During this time of year, the wood may seem empty until you hear some tiny, high-pitched contact calls by which these little birds stay in touch with each other, and signal the discovery of a new food to their fellow travellers.

There are six true **tits** in Britain, together with two related species. The 'true' tits include some of our more common garden birds, together with much scarcer species. The **Great**, **Blue** and **Coal** **Tits** are all common and widespread, and often visit gardens to supplement natural food sources, or to nest, often in artificial nestboxes. Their feeding behaviour delights many a home-owner, as they squabble with each other to get the best

The Coal Tit prefers coniferous and mixed woodlands, though it will also visit garden bird feeders, where it tends to be less bold than its relatives, often giving way if they appear.

position on the hanging feeder, rather than extract a morsel of peanut of energy-rich seed. Great Tits are the top dog in the hierarchy, though Blue Tits often sneak in cheekily under their beaks, as it were. Coal Tits tend to hang back and are a bit shyer.

Outside the garden, all three species are found in mixed woodland, with Coal Tits also having a liking for coniferous forest. In the woods they are joined by **Marsh** and **Willow Tits**, a sibling pair of species that look very alike. Marsh Tits also visit gardens from time to time, but mainly inhabit dry wooded areas, while Willow Tits prefer damp woods, often close to water. All these species may join tit flocks in winter, though Willow do so more rarely, and all nest in holes or cavities in trees. The final 'true' member of the family, the **Crested Tit**, is confined to the Scottish pine forests, though there it will behave in true tit

The Marsh Tit is a hole-nester, preferring damp deciduous woodland, where there are suitable rotten trees in which to nest.

Confined to the ancient Scottish pine forests, the Crested Tit is easily the rarest and most elusive member of the tit family. Nevertheless, in winter it will also come to bird feeders, just like its commoner relatives.

fashion, nesting in holes in trees and joining feeding flocks, even comes to artificial feeders where provided.

Another species, the **Long-tailed Tit**, comes from a different family, but to all intents and purposes, behaves like the other tits, especially when feeding. Long-tailed Tits often travel in flocks of up to a dozen or more birds, usually related to each other, which call constantly as they move acrobatically through the foliage. If you are patient and still, they will often come very close. Unlike other tits, they build their own nest out of moss and lichens, making it look like a ball with a small entrance hole. All these woodland tits will often respond to 'pishing', a noise that seems to arouse their curiosity and brings them closer.

The final British 'tit' is, in fact, a member of a tropical family known as 'parrotbills'. The **Bearded Tit** is a bird of reedbeds, from which comes its traditional name of 'reedling'. It is a beautiful and elegant bird, usually found by listening for its distinctive, metallic 'pinging' call. On a calm day you may get good views as birds climb to the tops of reeds; but if it is windy you don't stand much of a chance.

The **Nuthatch** and its relatives are unique; they are the only birds that can walk down as well as up a tree trunk. They do so by using their formidable claws. They are rather like a miniature woodpecker in appearance and habits, moving up and down tree trunks to find food, and nesting in holes.

The **Treecreeper** is often seen alongside Nuthatches and tits. It is an easily overlooked bird, creeping like a tiny mouse around branches and twigs to find insects, and only reluctantly flying to the next tree. It habitually climbs in spirals, going around the back of a trunk or branch, then re-appearing around the front again.

Like a small rodent, the Treecreeper climbs up and around the trunks and branches of trees in search of tiny insects on which to feed.

SHRIKES

Two species of shrike
are occasionally seen
in Britain. One is a former
breeder, now only seen as a
migrant; and another is a regular but very
scarce winter visitor. The **Red-backed Shrike**
earned the country name of 'butcher bird' from its
unpleasant, but extraordinary habit of impaling its prey on
thorns. Sadly, it has now been lost as a British breeding species,
probably due to a combination of modern farming methods
(which reduces the number of large insects), and climate
change, which brought wetter summers at the time of the
species' decline. It is, however, possible that global warming
will lead to a more
benevolent summer
climate for the species,
and it could return. Its
larger relative, the
Great Grey Shrike,
may be seen in winter
at regular haunts on
heaths in southern and
eastern Britain, though
this bird has also
declined in recent
years. It too sits on
high, prominent posts
and hunts for food by
diving down on small
mammals and birds.

*The Great Grey Shrike
is a rare winter visitor
to Britain, occasionally
seen on sandy heaths.
It often chooses a high
perch from which to
survey its winter
territory, making it
relatively easy to find.*

*Once a common and
widespread breeding
bird in Britain, the Red-
backed Shrike is now
only a rare passage
migrant in spring
and autumn. It
earned the folk-
name of 'butcher
bird' from its
habit of
impaling its
prey on thorns
in a 'larder'.*

147

STARLING

One of the commonest and most easily overlooked of British birds, the **Starling** is also one of our most intelligent and sociable birds, with complex breeding and flocking behaviour. In the breeding season males sing from perches on trees and especially roofs, delivering a song extraordinary for its mimicry – not just of other birds but also of car alarms, mobile phones and a host of other mechanical objects.

Outside the breeding season, roving flocks of Starlings feed in gardens (especially on open lawns and bird tables), playing fields and farmland, probing short grass for invertebrates. They also come together – especially in winter – at dusk, forming huge, noisy flocks. These were once a common sight in many urban areas, as well as the open countryside, but in recent years, the species appears to have declined, and fewer sites host such large flocks as they once did. If you do know a site, visit an hour or so before dusk, and marvel at the birds' behaviour as they swirl around in the sky, with small groups joining every minute or so, before they finally settle on a building or in the trees. It is an amazing sight.

Of all our common breeding birds, the Starling has the most extraordinary ability to mimic other sounds, including car alarms and telephones, as well as other birds' calls.

CROWS

The seven British breeding species of crow are among the most intelligent of all birds, and display a wide range of behaviour. Crows are especially curious birds, known for stealing bright, shiny objects, and also, rather less endearingly, for preying on other birds, as well as their chicks and eggs.

There are three species of large, black crow. The **Carrion Crow** is the classic member of the group, with its adaptability and omnivorous diet. The Carrion Crow is widespread in Britain apart from in the extreme north and west, where it is replaced by its distinctive race, the **Hooded Crow**. Both races are the duckers and divers of the bird world, always on the lookout for an opportunity, though this can involve muscling in on smaller birds. Both often gather in large flocks to feed or roost, and will also mob birds of prey without compunction.

The **Rook** is a gentler bird, usually found in more rural areas where it gathers to breed early in the year at rookeries. Rooks are also famous for their 'tumbling' behaviour during windy days in autumn, which is supposed to foretell unsettled weather.

The biggest crow of all is the **Raven**, a magnificent bird with glossy black plumage and a commanding stature. Ravens are a bird of the uplands, where birds will nest early in the year on prominent crags, from which they survey their little world. They are also supreme flyers, often courting high in the air.

TOP TIP

Studying crows
If you want to study social behaviour in birds, you could do a lot worse than to watch members of the crow family, whose complex social structures, and highly intelligent behaviour, make them one of the most interesting of all bird families.

Our commonest large crow, the Carrion Crow is a familiar sight as it scavenges for food. Crows will eat almost anything – a factor that has contributed to their success.

The Raven is the largest member of the crow family, confined to upland areas of Britain. It soars overhead on huge wings, often uttering its croaking call.

Jackdaws are highly sociable birds, whose behaviour is always worth watching, as they jostle with each other for food.

Jays are the most colourful members of the crow family and one of the most ingenious – though their shy habits mean they are not always easy to observe.

Two smaller 'black' crows species, the **Jackdaw** and the **Chough**, also have interesting traits. Jackdaws are real comedians, and often associate with their larger relatives in search of food on farmland or in open parkland. Their characteristic call is what gave them their name. Similarly, the call of their rarer relative, the Chough (originally pronounced 'chow') is echoed in the bird's name. Choughs are also comical birds to watch as they walk around on short grass on their huge red feet, poking their long, red, decurved bill into the soil to get at their food.

The remaining two members of the crow family are more colourful, or at least more strikingly patterned, than their mainly dark relatives. The **Jay** is a familiar, if rather shy, garden bird, which often hangs about in trees and bushes before swooping quickly down to take food from a bird table or a nest. In autumn, numbers are boosted by continental immigrants. The **Magpie** is often portrayed as a villain because it takes eggs and young chicks from songbirds' nests. Like all predators, though, its numbers are governed by its prey, not the other way round. Therefore, although such behaviour may seem cruel (the magpie is only doing what it must to feed its own young) it has no lasting effect on songbird populations. These are declining for other reasons. Magpies are sociable, usually travelling in pairs or small groups, which are supposed to bring the observer good luck. A lone bird, though, is said to bring bad luck.

SPARROWS AND BUNTINGS

These two families of seed-eating birds are closely related, and share many aspects of their behaviour with each other, so I have treated them together for the purposes of this book.

The two species of **sparrow** have both declined in recent years, in part due to modern farming methods. **House Sparrows** were once so common as to be virtually ignored, but since their decline, birders have begun to study them in greater depth. One possibility for their decline is that, being sociable birds, they need several pairs in an area to stimulate each other to breed. Because of falling numbers, there are simply not enough birds in some areas to sustain a viable breeding population. If this is the case, we may see the species suffer the same catastrophic decline as its more rural relative, the **Tree Sparrow**. The population of this species has plummeted by more than 90 per cent, again due to modern farming methods not leaving enough grain on the fields for the birds to eat in winter. Both species will join with other seed-eating birds outside the breeding season. Tree Sparrows mainly breed in holes in trees, while House Sparrows are more adaptable, nesting in holes in buildings as well as trees.

Five species of **bunting** regularly breed in Britain. Three are widespread, while the other two are confined to specialized areas

The classic 'little brown job', House Sparrows are, in fact, very social creatures, with a wide range of interesting behaviour, including dust-bathing, feeding and courtship.

of the north and south. The **Yellowhammer** is the most common and widespread of them all, although, as with all farmland species, it has declined in recent years. It is fairly easily seen, perching on hedgerows or bushes and singing its famous 'little-bit-of-bread-and-no-cheese' song. In winter, it joins forces with other buntings and finches to feed on stubble fields.

Its much rarer and more localized relative, the **Cirl Bunting**, is confined to south-west England, where it requires a specialized traditional farmland habitat all year round. Its behaviour is very similar to its more common relative.

The other common species of bunting are, to a degree, both farmland birds. The **Corn Bunting** is a common sight in some areas of lowland Britain, where it perches on wires or poles to sing its characteristic 'key rattling' song. However, it has also declined and may be absent from former haunts. Its breeding behaviour is notable in that males pair with several females, and have to work hard to keep away rivals. The **Reed Bunting**, as its name suggests, is more a bird of wetlands, though it also uses farmland to breed. Reed Buntings have begun to visit gardens in recent years, even feeding on bird tables.

The remaining British bunting is confined as a breeding bird to the high arctic-alpine habitat of the Scottish Highlands, but in winter, flocks gather around Britain's coasts. The **Snow Bunting** is a real specialist, able to breed farther north than any other small bird, though in winter it migrates south to milder climes. Winter flocks feed on shingle beaches and saltmarshes, appearing quite inconspicuous while feeding, but showing an explosion of white when they take off.

FINCHES

Finches are the world's largest family of birds, with more than 400 different species, yet just a dozen breed in Britain – and that includes the taxonomically dubious **Scottish Crossbill**, which may prove to be less than a full species.

The remaining 11 include some of our more common and best-known birds, of which the **Chaffinch** is the most widespread and numerous. Like other finches, it eats a wide range of seeds, though it also feeds its young on caterpillars. Chaffinches form large flocks outside the breeding season, often associating with other species, including its northern equivalent the **Brambling**. A rare breeding species here, Bramblings are abundant in northern Europe. Variable numbers arrive in Britain each autumn to seek out their favourite food of beechmast.

The other common finches include a generalist, the **Greenfinch**, and a specialist, the **Goldfinch**. Greenfinches eat a wide variety of seeds, and are particularly partial to artificial feeders containing either sunflower seeds or peanuts. As a result, they are frequent garden visitors, and often stay to breed, the male performing his attractive display flight, while singing as he flies. Goldfinches have a needle-sharp bill, ideal for prising the tiny seeds out of plants such as teasels, which they love. Listen out for the birds' tinkling calls as they fly overhead.

A much more recent garden visitor is the **Siskin**. Once confined to coniferous forests, mainly in northern Britain, the Siskin spread southwards a few decades ago and quickly adapted to feeding in gardens, where it is now a regular visitor. Siskins are lively little birds, which away from gardens, tend to flock together near water, where they often join **Redpolls** feeding on alder cones. Redpolls are less likely to visit gardens than Siskins,

Goldfinches are one of our most attractive common birds, and may be watched as they flit from plant to plant to gather seeds.

153

and are generally found in damp woods. In the breeding season, Siskins head north to breed in the pine forests of Norway and Scotland, while Redpolls breed on heathland.

Another 'pair' is formed by the **Linnet** and the **Twite**, sometimes known as the 'mountain Linnet' because of its preferred habitat. Linnets are birds of farmland and heath, flocking in autumn and winter and pairing up in spring when the male adopts his splendid breeding garb. Twite also form flocks in winter, heading away from their hilly breeding areas towards the coast, where they often join with Snow Buntings.

The three finches with the biggest bills and toughest appearance are the **Bullfinch**, **Hawfinch** and **Crossbill**. The Bullfinch uses its powerful bill to feed on fruit buds, though it also eats other berries and seeds. It generally appears in pairs or small family parties, and is a shy bird, often overlooked. But for shyness, the Hawfinch takes the prize: despite being our largest finch, with a bill so powerful it can crack cherry stones, it is hardly ever seen even when present in wooded habitat. Hawfinches are best looked for at known sites in winter, when they gather in small flocks and are easier to see as the trees are bare. The Crossbill (and its relatives) has a unique bill, with the tips of the mandibles crossed over to enable it to extract seeds from pinecones and other coniferous fruits. Crossbills are nomadic, and are the only species that migrates just once a year, moving in late summer to new feeding areas in different parts of Europe. In one year, there may be hundreds of birds, and in the next year none at all.

The Linnet was once kept in cages for the beauty of its song; now, thankfully, they fly free, delighting anyone lucky enough to see them. In winter, they form large flocks, searching for seeds along with other finches, sparrows and buntings.

One of our shyest small birds, Bullfinches are usually seen flying away, when their white rump is distinctive. In spring, they feed on fruit blossom, making them very unpopular with market gardeners.

GLOSSARY

aberrant Individual bird that shows an unusual or abnormal characteristic compared to others of its species: for example, an albino or melanistic bird.

adaptation Evolutionary development of a particular characteristic to fit a change in the environment: such as a change in migratory habits, or tendency to breed earlier in response to climate change.

alien A species that has been introduced into an environment from abroad, for example, Canada Geese, which originate in North America. Now usually known as 'introduced'.

altricial A bird that hatches in the nest and remains dependent on its parents for food and care, as with all passerine birds. Also known as nidiculous.

circadian Relating to the rhythms of the day: for example, the patterns of being awake or asleep.

diurnal Generally active during the day, as opposed to night (nocturnal).

eclipse The period during mid to late summer when ducks moult, and often become flightless for a short time.

feral A species or population that was originally kept or released by humans, but is now living freely in the wild.

fledging The point at which songbirds grow feathers, leave the nest, and are ready to fly.

habitat The distinctive place or area where creatures, such as birds, live: e.g., woodland or heathland. Characterized by distinctive vegetation and soil, and often also by climate.

incubation The period during which a bird sits on its eggs, in order to provide the warmth necessary for development of the embryo.

irruptive A population or species that periodically invades an area where it is not normally found, usually in order to find food.

jizz A birder's term for the 'general impression' given by a bird, even if it not possible for details of plumage and coloration to be seen.

lek Male breeding display ground where females come to choose a mate.

migratory A species or population that undertakes (usually twice-yearly) medium or long-distance movements between its breeding and wintering grounds.

morphology The external shape and form of a bird or other living organism.

moult The process of shedding old, worn feathers and regrowing new ones.

passerine One of the many families of perching birds; also known as songbirds.

polygamous A bird species in which either one male breeds with several females (polygyny); or one female breeds with several males (polyandry).

precocial A bird that is able to look after itself, see and walk or swim soon after hatching: e.g. ducks, gamebirds, waders (also known as nidifugous).

predator Any bird that hunts and kills another vertebrate (bird or mammal) in order to feed.

sedentary A bird that spends all, or most, of its life in the same area (as opposed to migratory).

subspecies A distinctive population of a particular species, which has diverged enough to be distinguishable, but is not yet so different as to be classified as a full species. Also known as race.

USEFUL ADDRESSES

The Wildlife Trusts
The Kiln, Waterside, Mather Road, Newark NG24 1WT
Tel: 0870 0367711
Fax: 0870 0360101
Email: info@wildlife-trusts.cix.co.uk
Web: www.wildlifetrusts.org
The Wildlife Trusts is the UK's leading voluntary
organization working on all aspects of wildlife protection
and people involvement. They manage almost
2,500 nature reserves throughout the UK, have more
than 411,000 members and receive support from over
22,000 volunteers every year. Members receive local
magazines and information, and the award-winning UK
magazine, *Natural World*, three times a year.

Wildlife Watch
(Contact details as above)
Email: watch@wildlife-trusts.cix.co.uk
Web: www.wildlife-watch.org
Wildlife Watch is the children's club and junior branch
of The Wildlife Trusts and has 29,000 supporters.
Members receive three magazines with articles and
ideas, three large posters with tips on wildlife watching
and the chance to join other children once a month to
discover local wildlife.

BTO (British Trust for Ornithology)
The National Centre for Ornithology, The Nunnery,
Thetford, Norfolk IP24 2PU
Tel: 01842 750050
Fax: 01842 750030
Email: general@bto.org
Web: www.bto.org
The BTO offers birdwatchers the opportunity to learn
more about birds by taking part in surveys such as the
Garden BirdWatch or the Nest Record Scheme.
BTO members also receive a bi-monthly magazine,
BTO News.

CJ Wildbird Foods Ltd
The Rea, Upton Magna, Shrewsbury SY4 4UB
Tel: 0800 731 2820 (Freephone)
Fax: 01743 709504
Email: enquiries@birdfood.co.uk
Web: www.birdfood.co.uk
CJ Wildbird Foods is Britain's leading supplier of
birdfeeders and foodstuffs, via mail order. The
company sponsors the BTO Garden BirdWatch survey
and also produce a free handbook of garden feeding,
containing advice on feeding garden birds, and a
catalogue of products.

Haith's
65 Park Street, Cleethorpes, Lincolnshire DN35 7NF
Tel: 0800 298 7054 (Freephone); Fax: 01472 242883
Email: sales@haiths.com; Web: www.haiths.com
Haith's produce a comprehensive, free mail order
catalogue offering a wide range of products to help
encourage birds and other wildlife to your garden. The
company supports the vital work of The Wildlife Trusts
across the UK.

RSPB (Royal Society for the Protection of Birds)
The Lodge, Sandy, Bedfordshire SG19 2DL
Tel: 01767 680551
Fax: 01767 692365
Email: bird@rspb.demon.co.uk
Web: www.rspb.org.uk
The RSPB is Britain's leading bird conservation
organization, with almost one million members. It runs
more than 100 bird reserves up and down the country,
and has a national network of members' groups.
Members receive four copies of *Birds* magazine each
year. The junior arm, the Wildlife Explorers, is for
members up to the age of 16.

Subbuteo Natural History Books Ltd
The Rea, Upton Magna, Shrewsbury SY4 4UB
Tel: 0870 010 9700
Fax: 0870 010 9699
Email: info@wildlifebooks.com
Web: www.wildlifebooks.com
Subbuteo Books provides a reliable mail order service for
books on birds and natural history. Free catalogue
available on request. With every purchase, a five per cent
donation is made to support The Wildlife Trusts.

Wildfowl and Wetlands Trust
Slimbridge, Gloucestershire GL2 7BT
Tel: 01453 891900
Fax: 01453 890827
Email: enquiries@wwt.org.uk
Web: www.wwt.org.uk
The Wildfowl and Wetlands Trust is the largest
international wetland conservation charity in the UK,
and is supported by over 90,000 members.

Wildsounds
Dept ABG, Cross Street, Salthouse, Norfolk NR25 7XH
Tel/Fax: 01263 741100
Email: isales@wildsounds.com
Web: www.wildsounds.co.uk
Wildsounds is Britain's leading supplier of birdsong
tapes and CDs. They also stock a range of Teach
Yourself products, particularly useful for the beginner.

FURTHER READING

MAGAZINES

BBC Wildlife
Available monthly from newsagents, or by subscription from:
BBC Wildlife Subscriptions, PO Box 279, Sittingbourne, Kent ME9 8DF
Tel: 01795 414718

Birdwatch
Available monthly from larger newsagents, or by subscription from:
Warners, West Street, Bourne, Lincolnshire PE10 9PH
Tel: 01778 392027

Bird Watching
Available monthly from larger newsagents, or by subscription from:
Bretton Court, Peterborough PE3 8DZ
Tel: 01733 264666

British Birds
Available monthly by subscription only from:
The Banks, Mountfield, Robertsbridge, East Sussex TN32 5JY
Tel: 01580 882039

GENERAL BIRD BEHAVIOUR
The Cambridge Encyclopaedia of Ornithology
edited by Michael Brooke and Tim Birkhead
Cambridge University Press, 1991, ISBN 0 52136 205 9

A Dictionary of Birds
edited by Bruce Campbell and Elizabeth Lack
Poyser, 1985, ISBN 0 85661 039 9

SPECIFIC BIRD BEHAVIOURS
*The British Trust for Ornithology
Migration Atlas*
John Marchant
Poyser, 2003, ISBN 085661 1379

Bird Migration
Thomas Alerstam
Cambridge University Press, 1993, ISBN 0 52144 822 0

Bird Migration
Peter Berthold
Oxford University Press, 2001, ISBN 0 19850 787 9

Bird Song
C. K. Catchpole and P. J. B. Slater
Cambridge University Press, 1995, ISBN 0 52141 799 6

Collins Field Guide: Bird Songs and Calls of Britain and Northern Europe
Geoff Sample
Collins, 1996, ISBN 0 00220 037 6

How Birds Migrate
Paul Kerlinger
Stackpole Books, USA, 1998, ISBN 0 81172 444 1

The Minds of Birds
Alexander Skutch
Texas A&M University Press, 1997, ISBN 0 89096 759 8

GENERAL BIRDS
Attracting Birds to Your Garden
Stephen Moss & David Cottridge
New Holland Publishers, 2000, ISBN 1 85974 005 7

Bill Oddie's Birds of Britain and Ireland
Bill Oddie
New Holland Publishers, 2002, ISBN 1 85368 488 0

Birdwatcher's Pocket Field Guide
Mark Golley
New Holland Publishers, 2003, ISBN 1 84330 119 9

The Garden Bird Handbook
Stephen Moss
New Holland Publishers, 2003, ISBN 1 84330 124 5

How to Birdwatch
Stephen Moss
New Holland Publishers, 2003, ISBN 1 84330 154 7

The New Atlas of Breeding Birds in Britain and Ireland
edited by Gibbons, Reid and Chapman
Academic Press, 1993, ISBN 0 85661 075 5

Where to Watch Birds in Britain and Ireland
David Tipling
New Holland Publishers, 2003, ISBN 1 84330 152 0

The Wildlife Trusts Guide to Birds
Series editor Nicholas Hammond
New Holland Publishers, 2002, ISBN 1 85974 958 5

INDEX

ACKNOWLEDGEMENTS

At New Holland, I would like to thank Lorna Sharrock and Gilly Cameron Cooper, for their editing skills, and Jo Hemmings, for commissioning the book in the first place! Also thanks to David Daly for his delightful illustrations, and David Tipling for his excellent and instructive photographs.

As someone who has watched birds for virtually the whole of my life, I should like to thank all my companions in the field – both long-term friends and casual acquaintances – who have prompted my interest in bird behaviour over the years. These include Daniel Osorio, Neil McKillop, Bill Oddie, Nigel Bean, Nigel Redman, Jackie Follett, Rod Standing and Graham Coster.

And as always, to my wife Suzanne, whose ability to see what I often miss has opened my eyes to a whole new world of birds and their habits.

Finally, I dedicate this book to my family in Italy: my father Franco, my stepmother Angela, my sisters Elisabetta and Arianna, and my grandmother Fiorina.

Publisher's Acknowledgements
All photography by David Tipling at Windrush Photos with the exception of the following:
Windrush Photos: Jari Peltomaki: page 44

All artwork by David Daly, with the exception of the following:
Clive Byers: pages 52, 64(t), 97(b), 98, 99(t), 100, 101(b)
Stuart Carter: 35
Stephen Message: pages 16, 41, 55, 73, 85, 86, 87, 89, 106, 107, 108, 109, 110, 111, 112, 115(b)

(t= top; b=bottom; c=centre; l=left; r=right)